Contracts With the Trades

Scope of Work Models for Home Builders

John Fredley
John Schaufelberger

Home Builder Press®
National Association of Home Builders
1201 15th Street, NW
Washington, DC 20005-2800
(800) 223-2665; fax (202) 822-0512

Check us out online at http://www.builderbooks.com

Contracts with the Trades: Scope of Work Models for Home Builders

ISBN 0-86718-436-1

© 1997 by the National Association of Home Builders of the United States

Cover by David Rhodes, Art Director, Home Builder Press

Printed in the United States of America

Library of Congress Cataloging in Publication Data

Fredley, John, 1941–
    Contracts with the trades: scope of work models for home builders / John Fredley, John Schaufelberger.
       p.   cm.
    Includes bibliographical references.
    ISBN 0-86718-436-1 (trade paper)
    1. Construction industry—Subcontracting—United States. 2. Construction contracts—United States. 3. Building trades—United States.
    I. Schaufelberger, John, 1942– . II. Title.
KF902.F74   1997
343.73′078624—dc21   97-28255
                             CIP

For more information please contact:

Home Builder Press®
National Association of Home Builders
1201 15th Street, NW
Washington, DC 20005-2800
(800) 223-2665
http://www.nahb.com/builderbooks

Additional copies of this publication are available from Home Builder Press. NAHB members receive a 20 percent member discount on publications purchased through Home Builder Press. Quantity discounts also are available.

9/97 ER/McNaughton 2,000

About the Authors

John Fredley teaches estimating, safety, and computer applications courses in the Department of Construction Management, University of Washington. Before teaching, he was a custom builder for 15 years and a member of NAHB, where he served as a state and national NAHB director and as president of his local chapter.

John Schaufelberger is a faculty member in the Department of Construction Management, University of Washington in Seattle. He worked in the construction industry for 30 years before joining the faculty and currently teaches courses in contracting and construction project management.

Acknowledgments

The authors and Home Builder Press wish to acknowledge the contributions of the following reviewers who assisted with reviews of the proposal or the developing manuscript: Allan Freedman, NAHB Builder Business Services, and members of the Single Family Production Builders Committee; Ed Caldeira, NAHB Research Center, Inc.; David Jaffe, NAHB Regulatory and Legal Affairs Department; Regina Solomon, NAHB Labor, Safety, and Health Department; William Young, NAHB Public Affairs Department; Bob Hanson, Builders Association of the Twin Cities; Peggy Lentz, Home Builders Association of Southwestern Missouri; Tom McCabe, Building Industry Association of Washington; Kathie Pugzczewski, Builders Association of Minnesota; Bill Wendle, Wisconsin Builders Association; John Barrows, J. Barrows, Inc.; Stephen Brooks, Grand Homes; Ken Donahue, Donahue Homes; David Keller, Keller Homes, Inc.; Steve McGee, Unify International; and John Piazza, Piazza Construction.

Contracts with the Trades: Scope of Work Models for Home Builders was produced under the general direction of Kent Colton, NAHB Executive Vice President/CEO, in association with staff members Jim Johnson, CIO and Staff Vice President, Information Services; Adrienne Ash, Assistant Staff Vice President, Information and Publications Services; Rosanne O'Connor, Director of Publications, Home Builder Press; Sharon Lamberton, Assistant Director of Publications and Project Editor; David Rhodes, Art Director; and Carolyn Kamara, Editorial Assistant.

Contents

Introduction 1

Chapter 1: Why Builders Use Trade Contractors 4
Advantages 4
Disadvantages 7
Manageable Challenges 7

Chapter 2: Writing Trade Contracts 8
Specify the Scope of Work 9
Determine the Contract Type 10
Define Terms and Conditions 11
Look at the Entire Contract 20

Chapter 3: Defining the Scope of Work 27
Why Are Scopes of Work Needed? 27
Define the Scope of Work 28
Write the Scope 28
Organize the Scope of Work 30
Develop Scopes of Work 31

Chapter 4: Steps Leading to the Trade Contract 62
Develop a Contracting Strategy 62
Select the Contract Scope of Work 63
Define Contract Terms and Conditions 65
Identify Potential Trade Contractors 65
Prequalify Potential Trade Contractors 65
Request Quotes from Trade Contractors 66
Estimate Cost for Each Contract Scope of Work 71
Gather and Evaluate Trade Contractor Quotes 71

Select Trade Contractors 73
Award the Trade Contracts 73

Chapter 5: Managing Trade Contractors 75
Build a Team 75
Control the Schedule 75
Monitor Quality Control 77
Control the Contract 78
Communicate with Contractors 84
Keep Job Diaries 88
Establish Safety Controls 89
Conduct Post-Construction Surveys 91
Require Coordinated Warranty Service 91
Project Success 92

Appendix 1: Blank Forms 93

Appendix 2: Diskette Information 117

Glossary 121

Related Publications 124

Figures

Figure 2-1.   Sample Trade Contract Document 21
Figure 4-1.   Sample Work Breakdown 63
Figure 4-2.   Sample Contracting Strategy 64
Figure 4-3.   Sample Trade Contractor Questionnaire 67
Figure 4-4.   Sample Request for Quotation 70
Figure 4-5.   Sample Quotation Evaluation Sheet 72
Figure 5-1.   Sample Project Schedule 76
Figure 5-2.   Sample Inspection Report Form 79
Figure 5-3.   Sample Partial Lien Release Form 80

Figure 5-4.   Sample Final Lien Release Form 81
Figure 5-5.   Sample Contract Change Order 82
Figure 5-6.   Sample Change Order Register 83
Figure 5-7.   Sample Preconstruction Meeting Agenda 85
Figure 5-8.   Sample Request for Information Form 86
Figure 5-9.   Sample Request for Information Log 87
Figure 5-10. Sample Document Control Register 88
Figure 5-11. Sample Daily Project Report 90
Figure 5-12. Excerpt from Construction Safety Checklist 91

Introduction

Today home builders in the United States use trade contractors to execute most of the specialized construction tasks involved in building a home. Indeed, builders typically contract out more of a project's scope of work than they perform with their own work crews. Trade contractors, therefore, are important members of the builder's team and can have a significant impact on a builder's success or failure. A single poorly performing trade contractor can disrupt the schedule for an entire project. Poor performance by trade contractors also can contribute to higher costs through rework and affect the real or perceived overall quality of the project.

Because trade contractors play such a central role in residential construction, builders find it pays to develop a long-term approach to contracting with the trades. This approach involves developing and nurturing positive relationships with reliable trade contractors and helping to ensure their financial health by treating them fairly. Successful contracting helps both builder and trade contractors deliver quality projects within cost, on time, and at a profit.

Although most builders need 10 to 20 different trades to construct a home, it seems their busy schedules often prevent them from writing trade contracts or specific scopes of work. Instead, they establish an understanding of what a trade can do with a phone call. After a few minutes the builder and trade contractor somehow agree on what will be done and the cost. A verbal or handshake agreement is all that holds the relationship together.

This way of doing business may be common and convenient, but it can have a serious downside: understandings easily get lost in translation and expectations can become distorted. Often requirements are vague or omitted. Inappropriate or insufficient materials may be ordered. The result is lost time and in some cases work must be redone. In more severe cases, the relationship between builder and trade contractor is damaged or may terminate.

This book was written to help builders develop and use trade contracts that create a clear, precise, and mutual understanding with the trades about expected performance from project to project. The major challenges home builders face include:

- Finding qualified trade contractors
- Defining a well understood and exact scope of work
- Negotiating a fair price for the work to be performed
- Scheduling and coordinating trade contractor work for timely completion of the project

- Ensuring a quality product
- Developing enduring relationships with reliable trade contractors

Builders who successfully contract with the trades usually have management systems that allow them to consistently meet these challenges. This book provides builders with a guide for using contracts as part of an effective management system. Sample trade contracts and scopes of work are provided, the how-tos of prequalifying and selecting trade contractors are discussed, and trade contractor management tools are described.

While some builders still operate on a handshake basis, others use very sophisticated legal documents. We recommend that you use written documents for all contractual relationships to formalize each agreement and minimize misunderstandings. However, there is no such thing as a "one-size-fits-all" contract. Plan to tailor the text of each contract to meet your needs. The complexity of the contract document depends on your relationship with the trade contractor and on the complexity of the project. While we recommend that some written provisions be included in all contract documents, others may be inserted at your option.

Success in home building depends on a builder's ability to select qualified trade contractors, define specific scopes of work, and effectively manage the scheduling and work quality of trade contractors. This book was written with the intent of helping builders become successful managers of trade contractors. The information in this book can benefit anyone in the residential building industry: a builder building a first custom home, an experienced builder seeking more information about contracts, or a trade contractor who is presented with a contract by a builder.

The advantages and disadvantages of trade contracting are discussed in Chapter 1. For experienced builders who already have a strong sense of the advantages and disadvantages of subcontracting a quick scan of this chapter may suggest some additional items to consider. Chapter 2 describes the basic elements of a trade contract including the issues to be considered and how to develop a sample contract. Builders who already use trade contracts can check their own documents against the framework presented in this chapter. For example: If you have instituted a jobsite safety program you will want to be sure appropriate references to your company's safety policies and procedures appear in your contracts. Full treatment of jobsite safety and OSHA compliance is beyond the scope of this book; however, the sample material presented includes illustrations of how appropriate references to safety requirements can be made within contract documents. Chapter 3 provides sample scopes of work builders can use to develop trade contracts covering more than 30 construction trades. Chapter 4 describes the selection process and provides a sample questionnaire for prequalifying trade contractors. Chapter 5 discusses how to manage trade contractors and provides sample forms you can use to develop your own management systems.

Please note that the sample contract language and forms contained in this book and diskette are illustrations. The material is provided as a convenience to builders who wish to incorporate applicable language into their own contract documents; however, this material should not be used without the review and approval of an attorney experienced in construction contract law. The suggested contract provisions do not and cannot apply to every situation; applicable law differs from state to state and local municipal law may also apply. The model language provided is only a starting point. Builders should have their attorneys prepare specific documents that meet their particular needs.

Chapter 1

Why Builders Use Trade Contractors

Historically homes were built by craftspeople who performed much of the work themselves or used their own crews supplemented by a few hired specialists. Labor costs were based on hours worked, with the builder providing all the materials, tools, and equipment needed to complete the project. Because residential construction has become more specialized, the majority of the actual construction work is now contracted to trade or specialty contractors. Today builders are more managers than craftspeople.

The basic agreement between the builder and the homeowner or home buyer may be a real estate purchase contract or a construction contract that specifies the exact work to be done, the performance time, and a mutually agreed-on cost.* The homeowner's contract is with the builder, not the trade contractors. Although the builder usually engages trade contractors to undertake specific parts of the work required to fulfill his or her contract with the owner, the builder remains responsible for ensuring that all work conforms to the project plans and specifications, including the specified completion date if there is one.

Selecting trade contractors, scheduling trade work, and coordinating the tasks of the various trades are jobs builders must manage well to be successful. Consider both the advantages and disadvantages of contracting with the trades as you develop your trade contracting strategies for individual projects.

Advantages

The use of trade contractors offers several advantages to home builders. The practice allows builders to reduce their business risk and provides needed flexibility in project execution. Trade contractors also provide builders with the necessary resources, experience, and skills to perform particular types of work.

Reduced Risks

One of the business risks associated with residential construction is accurately forecasting the amount and cost of labor required to complete a project. By contracting significant segments of the work to trade contrac-

---

Contracts and Liability for Builders and Remodelers, by David S. Jaffe, published by the Home Builder Press, provides an excellent discussion about how to develop construction contracts and the legal principles involved in construction contracting. We encourage you to refer to this book for assistance in preparing basic construction contract documents that can be used between builders and owners.

tors, builders can reduce this business risk by passing much—though not all—of the responsibility for costs to the trade contractors.

When the builder asks a trade contractor for a price to perform a detailed scope of work, the trade contractor then bears the risk of properly estimating the labor, equipment, and possibly material costs depending on the scope of work. The builder, however, retains the responsibility for ensuring the quality of the trade contractor's work.

Since trade contractors generally have their own management organization, builders only need to clearly specify the tasks to be performed. Direct jobsite supervision of the individual craftspeople is the trade contractor's responsibility, although the builder is ultimately liable for the quality of the completed work. The cost of direct supervision is included in the trade contractor's price.

Increased Flexibility

Trade contracting provides builders greater flexibility in selecting types of projects that occur in residential construction. Builders don't need to retain a large work force or a large investment in equipment in order to compete successfully. Since residential construction tends to be cyclical, builders can add or reduce the number of trade contractors to meet market conditions. What builders do need, however, are skilled managers and reliable, qualified trade contractors.

Access to Specialized Skills

Modern residential construction requires specialized skills for many construction tasks such as masonry, electrical work, or heating and air conditioning. Craftspeople experienced in these specialized trades are expensive and require special licenses and training. Unless a builder has a continuous need for such specialized labor, it is cost-prohibitive to employ them as part of his or her work force.

For most builders, the solution is trade contracting, through which builders purchase needed expertise only as required to complete specialized tasks. The result is cost-efficient access to a technically proficient pool of workers and supervisors.

Efficient Scheduling of Labor

Daily labor requirements are not constant throughout the construction cycle for residential projects. Seasonal and other fluctuations in volume and the type of construction work make many builders reluctant to maintain large crews. When work picks up, builders usually find it more efficient to contract with additional independent trade contractors than to add full- or part-time employees. Trade contracting helps contain the builder's administrative and tax burdens and provides the flexibility to add or subtract labor according to the specific projects under contract at any given time. Trade contractors also provide supervisors for their work crews, which frees builders from employing these specialized work leaders.

Reduced Need for Tools and Equipment

The specialized tools and equipment required for many construction tasks often require a significant investment. Since these tools usually are used only for a few phases of the overall project, builders generally have difficulty amortizing the investment while remaining cost-competitive. Their choices, therefore, are renting the needed equipment or contracting the work. This is another reason to contract selected specialized tasks to firms that can more readily amortize their investment in specialized tools and equipment.

By contracting with the trades, builders gain access to the specialized tools and equipment at a more reasonable cost per project.

Efficient Project Execution

Successful trade contractors generally develop efficient ways of performing their specialized work for the least cost and in the shortest amount of time. Therefore, using trade contractors should lead to a more efficient project execution. Such efficiencies are realized, however, only when the builder does a good job of scheduling and coordinating the various trade contractors.

Improved Materials Purchasing and Storage

Some trade contractors can provide all the materials needed for the contracted tasks. When this is the case, it relieves the builder from having to determine the required quantities, obtain them, and then deal with possible shortages or leftover materials at the jobsite. Trade contractors typically negotiate directly with suppliers for needed construction materials, arrange for their delivery to the construction site, and ensure they are properly stored and secured until incorporated into the project. If, as a builder, you maintain responsibility for obtaining materials and supplies, talk to your trade contractors in advance and ask them to help you determine the materials and amounts you will need. If your trade contractors handle materials purchasing be sure that your policies regarding jobsite storage and security are clear. Also, be sure to coordinate carefully if crews from more than one trade contractor will be working concurrently on the jobsite.

Lowered Costs

Provided the builder efficiently schedules and coordinates the work of the trade contractors, the overall gain in efficiency can result in lower total project costs. The trade contractor's skill and quality control procedures are crucial to achieving lower project costs. Trade contractors usually can purchase materials for less than builders can because they purchase larger volumes from their suppliers. Fewer employees and supervisors; reduced job overhead, equipment rentals, and cash outlays for materials; and a faster turnover of working capital are examples of how contracting specialized tasks can lower costs.

Disadvantages

Offsetting the positive aspects of trade contracting are some negatives.

- Builders tend to give up some control when working with trade contractors. The scope and terms of the contract define the responsibilities of each trade contractor. If some aspect of the work is inadvertently omitted, the builder is still responsible for making sure the specifications are met. For this reason, it is important that you always use a well-written trade contract and do not rely on verbal or "handshake" agreements.
- Consistent quality control also can be more difficult to achieve with trade contractors than with your own crews. Clearly state quality requirements during contract negotiations.
- The profit that trade contractors include in their bids or quotes may erode or even offset your gains in efficiency. We recommend that you solicit competitive bids from two or more trade contractors to ensure you receive the best price.
- Selecting trade contractors is time consuming. When it is not done carefully, undesirable results such as contract termination, hard feelings, legal fees, and project delays can occur.

Manageable Challenges

Many of the disadvantages of working with trade contractors can be overcome if you carefully plan the work, prequalify contractors, use well written contract documents, and solicit quotes from several qualified trade contractors for each contracted task.

Remember that any ambiguity in the contract requirements or a misunderstanding of the scope of work between the builder and the trade contractor can lead to problems in project execution and increased costs. Use clear, concise, well-written trade contracts to eliminate most of these problems. Create a project working environment that stresses teamwork and cooperation and you will realize the major advantages of trade contracting. Maintaining a good construction schedule and carefully coordinating all trade contractors are essential elements to successful project execution.

The steps builders take to select and work with trade contractors are discussed in more detail in Chapters 4 and 5.

In the next chapter, we will examine the major components of a trade contract and some of the issues that should be addressed in each section of the document.

Chapter 2

Writing Trade Contracts

Trade contracts usually are based on the provisions and specifications contained in the general contract between the builder and homeowners. For projects built on speculation, contracts are based on the builder's plans and specifications. In each case, the purpose of a written trade contract is to formalize the relationship between the builder and the contractor and to:

- Document the agreement
- Minimize misunderstandings
- Allocate risks

Before you write a contract you must specify the exact scope of work to be performed by the trade contractor and define the terms and conditions of the contract. The scope of work defines the specific tasks to be performed. Terms and conditions establish management requirements or procedures that must be followed by the contractor when performing the scope of work.

Any deficiencies or mistakes in the contract can result in legal disputes between the builder and the trade contractor. Because such disputes often result in additional costs to the builder, it is to your advantage to try to avoid misunderstandings by developing and using a well written trade contract. The document should provide a clear description of all work requirements so that you and your trade contractor understand the performance requirements. Write trade contracts in simple language that documents the exact terms of the agreement.

Before you write trade contracts, divide the building project into separate *work packages* and determine which ones will be executed by your crews and which will be contracted. A work package is a definable segment of the work required to complete the project. Examples of work packages for a residential project are sitework, landscaping, framing, concrete foundation, roofing, electrical, and plumbing.

The size of each work package depends on your scope of work and the complexity of the overall project. Packages can include multiple project tasks or activities. Your goal is to ensure that there is no overlap between the work packages and that no elements of work are omitted. Perform this project analysis early in the project planning process and before you develop the project cost estimate.

Estimate costs for contracted work by selecting the best from the quotes received from several trade contractors for each potential contract. Prepare your own cost estimate for each scope of work and then use it to determine the reasonableness of the trade contractors' quotes. Quotes that vary greatly from your cost estimate often mean that the trade contractors did not understand the scope of work in your request for a quote. Question the trade contractors and then rewrite the scope of work and solicit new quotes from the trades.

We develop a sample trade contract in this chapter. Each topic is addressed separately and sample contract text is presented. The suggested contract provisions will not apply to all situations. Make sure your contracts adhere to the applicable laws in your state and municipality. Our discussion is intended to identify certain issues, highlight some solutions, and point out potential situations that need to be considered carefully.

The entire sample trade contract is shown in Figure 2-1 at the end of the chapter and duplicated on the diskette that comes with the book. Where **[Builder]** is shown in the contract language, you should insert your company name. The term *Subcontractor* may or may not be replaced by the names of specific companies, depending on how you wish to tailor your contracts. Have your attorney prepare your specific trade contract language to address local requirements and to ensure that the language meets your company's needs.

Specify the Scope of Work

The trade contract must clearly state the exact scope of work to be performed and include relevant specifications and drawings that are to be used. Use the project drawings and specifications prepared by the architect to prepare the scope of work for each trade contract. Each scope of work usually consists of a single work package or a combination of work packages. All or part of the general contract between the builder and homeowner may be referred to in the trade contract scope of work. If the builder is constructing the project on speculation, all appropriate drawings and specifications must be included in each trade contract.

Any jobsite management tasks the builder wants to have performed by the trade contractor must be specified in the contract. Examples of such tasks include meeting all applicable health and safety requirements; installing a protective surface over the trade contractor's completed work; and removing all construction waste generated by the contractor's crew.

A well-defined scope of work is essential to ensuring that potential trade contractors understand what is expected of them. Poorly defined scopes of work lead to conflicts and often result in cost escalation, time delays, or litigation. When possible, consult with trade contractors while developing the scope of work to ensure complete understanding and agreement regarding the trade contractor's responsibilities. Reviewing the plans and specifications for the project with prospective trade contractors before you write detailed scopes of work might reduce the potential for problems during construction and result in more competitive quotes. Consulting with trade contractors in advance also reinforces the concept that the trades are part of your team.

The first step in writing a trade contract is selecting the specific scope of work to be performed by the trade contractor. Guidance on the writing of scopes of work for the various trades is discussed in detail in Chapter 3.

As stated earlier, scopes of work are developed by dividing the entire project into work packages. The work packages become the building blocks that enable the project to be completed. For example, let's take the electrical work required for the Smith residence. It involves two phases: rough-in and finish work. The builder has a general contract for the construction of the residence and has decided to contract all of the electrical work to a single electrical contractor. The following scope of work was established for the electrical contract:

> Provide all materials, labor, tools, equipment, supervision, supplies, and other items necessary or required to execute the work and perform all electrical work in accordance with the contract drawings, specifications, and any addenda contained in the general contract documents for the construction of the Smith residence. The intent is to provide a complete electrical system that meets local code requirements including, but not limited to, provision and installation of the following:
>
> 1. Primary and secondary conductors, conduits, and grounding
> 2. All panels, subpanels, breakers, and switches for each system including circuits for heating, air conditioning, appliances, water heater, lighting, GFCI, receptacles, switches, and accessories
> 3. Light fixtures, fans, range hood, smoke detectors, bulbs, and lamps
> 4. Color coding and marking, as required
> 5. Cutting, coring, and patching, as required
> 6. Testing and hot-checking of entire electrical system

Determine the Contract Type

A contract amount and the basis for the amount must be specified in the agreement section of the written contract. The contract may be awarded on a *lump sum, unit price,* or *cost-plus* basis. The nature of the work being contracted usually dictates the basis for the pricing of the trade contract.

- Use lump sum contracts when the exact scope of work is well defined and can be quantified.

- Use unit price contracts when the scope of work can be defined, but the exact quantities of materials or some of the elements of work are unknown. For example, estimated quantities might be provided for each unknown element of work in the request for quote and the trade contractors submit a unit price for each work element.
- Use cost-plus contracts when the scope of work is poorly defined. In this type of contract, trade contractors are reimbursed their actual costs plus paid a fee. Such contracts are sometimes called time and materials contracts.

Most residential trade contracts are based on a lump sum or a unit price basis. A more complex project may be awarded on a cost-plus basis, however. In some cases, work that is difficult to define might be broken down into small segments with some portions being lump sum and others being unit price to avoid the necessity for using a cost-plus contract. For example, a combination unit price and lump sum contract often is used for sitework because the exact amount of backfill is not known. Concrete paving often is priced in terms of dollars per square yard of concrete.

If the exact scope is unknown and you are willing to share some of the risk inherent in such situations, you might consider using a cost-plus contract.

In our sample contract for the electrical work for the Smith residence, we used a lump sum contract because the requirements and scope of work are well defined. The following text is used:

> **[Builder]** agrees to pay and Subcontractor agrees to receive and accept as full compensation for performing all work according to the requirements of the contract and furnishing all materials, supplies, and equipment required to execute the work, the sum of _____ dollars ($_____) subject to additions and deductions from such changes in the scope of work as agreed upon in writing through change orders to the contract.

Define Terms and Conditions

The terms and conditions section of the contract establishes the operating procedures the builder intends to use to manage the contract and might assign management requirements to the trade contractor. Tailor this section of your contracts to meet your specific requirements. At a minimum, the following topics should be addressed in this section:

- General responsibilities
- Commencement and progress of work
- Payment procedures
- Laws and regulations
- Insurance
- Indemnification

- Safety
- Inspection and acceptance of completed work
- Change order procedures
- Dispute resolution
- Trade contractor's status as independent contractor
- Warranty
- Termination or suspension

Each topic is addressed in the following sections in the context of the electrical contract for the Smith residence.

Articulate General Responsibilities

Specify all legal obligations imposed on the trade contractor in the trade contract or include them by referencing the applicable sections of the general contract for the project, if one exists. Examples include warranty requirements, shop drawing requirements and processing procedures, site access restrictions, schedule requirements, and site cleanup and disposal requirements. Requirements not incorporated by reference need to be explicit so there is no confusion regarding builder expectations.

Requirements imposed by this section generally add to the trade contract cost. Carefully select only essential requirements to minimize cost impacts and then clearly state those requirements.

The general conditions of the general residential contract usually are incorporated by reference in the trade contract, if such a contract exists. Otherwise, all general conditions must be clearly stated in the trade contract. Any requirements the builder desires to add to those contained in the general contract must be listed in the trade contract.

Check for consistency between the trade contracts and the general contract. Builders should not promise homeowners more than they require from the trades. In the case of the electrical contract for the Smith residence, we identified the following general work requirements:

1. Subcontractor shall comply with the General Conditions of the general contract for the construction of the Smith residence and any interpretations as to the meaning thereof issued by the Owner.

2. Subcontractor shall ensure that all wiring and electrical items fully comply with code requirements that are effective on the date contract is signed. Code modifications made after contract has been signed are not included.

3. Subcontractor shall visit the construction site before starting work to understand access restrictions to the site.

4. Subcontractor shall coordinate the installation of all electrical work with other interfacing trades to ensure an unencumbered worksite.

5. Subcontractor shall coordinate connection with the local power utility.

6. Subcontractor shall coordinate with the local cable and telephone utilities to ensure working television and telephone outlets as located on the plans.

7. Subcontractor shall provide a qualified onsite supervisor whenever work is being performed. Supervisor will attend weekly coordination meetings to review scheduled versus actual progress, quality control, coordination of work, and jobsite safety.

8. Subcontractor shall submit required shop drawings and manufacturers' cut sheets for all equipment and await receipt of approval of Owner or Architect prior to starting work.

9. Subcontractor shall at all times keep the project site clean of dirt, debris, trash, and any waste materials arising from the performance of the subcontract. Subcontractor is responsible for removal of all debris created as a result of the work being performed and dispose at a site designated by **[Builder]**.

Performance and payment bonds generally are not required for the construction of individual homes. Thus no bond requirement was included in this terms and conditions section. If performance and payment bonds are required or desired, the following requirement could be added to the end of this section:

10. Subcontractor shall provide performance and payment bonds with value equal to the subcontracted amount.

Relate the Contract to the Project Schedule

The builder must integrate and schedule the trade contractors' work into a master schedule for the entire project. This schedule, whether in the head of the builder or maintained on a computer, should be updated periodically and used to notify trade contractors regarding the scheduled start and completion dates for their tasks. The contract needs to link the trade contractors' performance to this master construction schedule.

It is also important to specify in the contract a time frame within which the trade contractor must respond after being notified to proceed with the work. To ensure the efficient completion of the trade contractor's work, the builder must make sure that the project site is ready for the trade contractor to start work on the date scheduled. False starts will cost the builder money because the trade contractors will arrive on site and not be able to work.

In our sample contract for the Smith residence, we used the following text to link trade contractor performance to the overall construction schedule for the project:

1. Subcontractor agrees to comply with and perform the subcontracted work in conformance with project plans and specifications and applicable state, county, and municipal codes, ordinances, and statutes according to the requirements of the **[Builder's]** construction schedule or schedules as **[Builder]** may from time to time develop and submit to Subcontractor. Within three (3) calendar days after being notified by **[Builder]**, Subcontractor shall commence actual construction on such parts of the scope of work as **[Builder]** may designate and to thereafter continue diligently in the performance of the work. All work shall be performed in full cooperation with **[Builder]** and other subcontractors.

2. Upon request, Subcontractor shall prepare and submit to **[Builder]** for approval a progress schedule to meet the dates as shown by **[Builder's]** then current construction schedule and showing the order in which Subcontractor proposes to carry on the work and the dates on which it will start and complete salient features of the subcontracted work.

3. If in the opinion of **[Builder]** Subcontractor falls behind the progress schedule, Subcontractor shall take steps as may be necessary to improve the subcontract progress, and **[Builder]** may require Subcontractor to increase the number of shifts and/or overtime work, days of work, and/or increase equipment and/or tools being used, and to submit such revised schedule demonstrating the manner in which the agreed rate of progress will be regained.

We recommend that text such as this be included in most trade contracts to avoid any misunderstandings between the builder and the trade contractor.

Establish Payment Procedures

The method of payment and schedule for payment requests also should be specified in the trade contract. The procedures must be completely understood by all parties and should include specific formats to be followed and time frames for submission. The trade contractor should be required to provide the builder with a written invoice for the total value of work completed and a statement that all work conforms to the project plans and specifications.

Payment procedures may simply require the trade contractor to submit a bill when the work is completed and the builder to pay the bill within a specified time period. More complicated arrangements may involve payments of agreed-upon amounts upon completion of each phase of work or according to a schedule of values. Ideally payment procedures should be kept as simple as possible. Retainage amounts, if any, must be identified and the timing of retainage releases clearly stated.

We used the following clauses to establish the basis for progress payments to the electrical contractor and to describe the procedures for requesting payment:

1. Payment will be made to Subcontractor for work performed under this subcontract as measured and certified as being completed by **[Builder]**. Before any payment is due, Subcontractor must provide **[Builder]** with a written estimate of the total amount of work completed and a signed statement that all completed work fully conforms to contract requirements.

2. Upon application, partial payments for work performed under this subcontract will be made by **[Builder]**, and will equal the value of the work performed by Subcontractor, less the sum of previous payments.

3. Subcontractor will receive partial payments fifteen (15) days after invoice has been received by **[Builder]**.

4. Upon completion of all contracted work, Subcontractor will be paid the remaining amount due Subcontractor under this subcontract. Final payment shall release **[Builder]** from any further obligations whatsoever in respect to this subcontract. Subcontractor shall, as a condition to receipt of final payment, provide to **[Builder]** a full release from any and all claims, liens, and demands whatsoever for all matters growing out of, or in any matter connected with this subcontract.

Lien release forms are discussed in Chapter 5.

The terms and conditions in this section of our sample trade contract make no provision for retainage. If retainage is desired, clause 2 above needs to be modified to provide for whatever retainage percentage is to be used.

Require Compliance with Laws and Regulations

Trade contracts should contain a clause requiring contractors to comply with all relevant laws and regulations and to obtain any needed permits and inspections required to perform the work contained in the contracted scope of work. Failure to clearly articulate this responsibility may lead to disputes with the trade contractor and place additional risks on the builder.

The following clause was used to notify the electrical contractor in our example that he or she is required to obtain any permits and inspections required for his or her work and otherwise comply with all relevant laws and regulations:

> Subcontractor, its employees, and representatives shall at all times comply with all applicable laws, ordinances, statutes, rules, and regulations, federal and state, county and municipal, and particularly those relating to wages, hours, fair employment practices, non-discrimination, and working conditions. Subcontractor shall procure and pay for all permits, licenses, and inspections required by any governmental authority for any part of the work under this subcontract and shall furnish any bonds, security, or deposits required by such authority to permit performance of the work.

Require Insurance

Like the builder, the trade contractor should be required to obtain liability insurance, including automobile and property damage insurance, and workers'compensation when required by law or the builder's insurance company. Require proof of such insurance before you allow the trade contractor to commence work. You also should require advance notification from the contractor's insurance company of any coverage cancellation. This is essential to reducing your liability for trade contractor actions and omissions. Builders are advised to have their insurance agents craft the exact text to be used in establishing insurance requirements in trade contracts.

The following clauses are used in our example to ensure that the electrical contractor is aware of his or her insurance requirements and to minimize the builder's risk:

> 1. Subcontractor, at its own expense, shall procure, carry, and maintain on all of its operations workers' compensation and employer's liability insurance covering all of its employees, public liability and property damage insurance, and automotive public liability and property damage insurance. Coverage limits shall be in accordance with the requirements of the general contract. Subcontractor is required to name [**Builder**] and Owner as additional insureds on Subcontractor's general liability policy.

2. Subcontractor shall provide to **[Builder]** prior to commencement of work a certificate from the insurance companies that such insurance is in force and will not be canceled without thirty (30) days written notice to **[Builder]** .

Indemnify the Builder and Homeowners

Trade contractors should be required to indemnify both the builder and homeowners against any actions or omissions taken by trade contractors or their agents. This allows builders to reduce their liability or risk on projects by allocating the risk of loss directly to the party responsible for the loss, which in this case is the trade contractor. For example, if the builder must repair defective trade contractor work, an indemnification clause entitles the builder to recover the cost of the repair from the contractor. Builders are advised to consult with their insurance agents when writing this clause for their trade contracts.

The following clause is used in our example to indemnify the builder and homeowners against actions or inaction on the part of the electrical contractor or its agents:

Subcontractor shall indemnify and hold harmless **[Builder]** and Owner against any claims, damages, losses, and expenses, including legal fees, arising out of or resulting from performance of subcontracted work to the extent caused in whole or in part by the Subcontractor or anyone directly or indirectly employed by the Subcontractor.

Enforce Safety Procedures

Safety is a crucial element of any construction project and needs to be highlighted in all trade contracts. Although the builder is ultimately responsible for the safety of all workers on the site, appropriate responsibility for jobsite safety must be clearly placed on trade contractors for their portions of the work.

Appropriate responsibility for compliance with all applicable federal, state, and local safety and health rules and regulations also should be placed on the trade contractors. This is very important in the event a citation is issued to the builder for a safety violation caused by a trade contractor.

To reinforce the importance of a good safety program in our example, the following clauses are used:

1. Subcontractor and all of its employees shall follow all applicable safety and health laws and requirements pertaining to its work and the conduct thereof, but not limited to, compliance with all applicable laws, ordinances, rules, regulations, and orders issued by a public authority, whether federal, state, or local, the Federal Occupational Safety and Health Administration, and any safety measures required by **[Builder]** or Owner.

2. Safety of Subcontractor's employees, whether or not in common work areas, is the responsibility of Subcontractor.

3. Subcontractor agrees to instruct all its employees to inform **[Builder]** immediately of any unsafe condition or practice whether or not in common work areas.

For larger projects, require a detailed job-specific safety plan from the trade contractor that identifies potential hazards and actions to be taken to reduce the potential for accidents.

References to safety requirements in contracts will not of themselves protect a builder against potential OSHA citations (or in court, should an accident occur). However, including and enforcing appropriate contractual requirements is part of a systematic approach to jobsite safety that can reduce the likelihood of problems.

Detail Inspections and Acceptance of Completed Work

The builder's right to inspect and accept completed work should be specified in the trade contract. The builder must retain this right to ensure conformity with contract drawings and specifications and to ensure acceptable workmanship. This is important to the builder's ability to ensure proper quality standards are met. The homeowner's right to inspect completed work, if the homeowner has a construction contract with the builder or the architect's right to inspect on behalf of the owner, should be included by reference to the appropriate provision of the general contract, if one exists. The right to inspect and accept completed work is essential to an effective quality control program.

The following clause is used in our example of the Smith residence for this purpose:

> The materials and work shall at all times be subject to inspection by Owner and **[Builder]**. Owner and **[Builder]** shall be afforded full and free access to the construction site for the purpose of inspection and to determine the general progress of the work. In the event that any of the work or materials are found to be improper or defective by Owner or **[Builder]**, Subcontractor shall upon notification in writing from **[Builder]** proceed to replace or correct the defective material or workmanship at its own cost and expense. If Subcontractor fails to correct the defective work, **[Builder]**, at its option, may replace and correct the same and deduct the cost of correcting the defective work from Subcontractor's payments.

Establish Change Order Procedures

A change order clause should be included in the trade contract to allow the builder to modify the scope of work as required to complete the project to the satisfaction of the owner, buyer, or to resolve design problems. Such clauses require the trade contractor to perform changes in the scope of work subject to an equitable adjustment in contract price. To be legally binding, any requested changes should be within the scope of the basic contract. For example, the builder could not issue a change order to the electrical contract for installation of additional doors.

The procedures to be followed in making changes to the trade contract must be clearly specified in the document. The trade contractor must understand who in the builder's organization has the authority to modify the contracted scope of work and the procedures to be followed for an equitable adjustment to the contract price and timeline.

Never allow trade contractors to accept changes directly from homeowners. The homeowners must make changes through the builder who in turn forwards approved changes to trade contractors. The contract also must state how the trade contractor issues notice to the builder regarding conditions believed to constitute changes to the scope of work in the original contract. Trade contractors should be required to issue notice to the builder before starting the additional work to allow the builder time to mitigate any cost and time impacts.

We used the following clause in our example to describe the change order procedures for the electrical contract for the Smith residence:

> **[Builder]** may order additional work, and Subcontractor will perform such changes in the work as directed in writing. Any change or adjustment to the subcontract price as a result of changes in the scope of work shall be as specifically stated in the change order. If Subcontractor encounters conditions it considers different from those described in subcontract documents or plans, it is required to issue written notice to **[Builder]** before proceeding. Subcontractor's failure to issue notice shall constitute waiver of any claims for additional compensation. Only **[Builder's]** project superintendent is authorized to issue change orders to Subcontractor. If Subcontractor and **[Builder]** cannot agree upon a price for the changes in the work, **[Builder]** may direct Subcontractor to execute the changes, and Subcontractor will be paid based on the actual cost to Subcontractor plus a reasonable markup for profit and overhead expenses.

Provide for Resolution of Disputes

Disputes involving either the homeowner or builder and the trade contractor may occur. The trade contractor's recourse for disputes resulting from the homeowners' actions or omissions is through the builder pursuant to the dispute provisions contained in the general contract. Trade contractor disputes that result from homeowner omissions usually relate to construction problems caused by defective design.

Procedures for resolving disputes between the trade contractor and the builder should be described in the trade contract. Claims against the builder, because of improper site management or any other reason not related to the homeowner, usually are handled through arbitration if the parties to the dispute are unable to resolve the issues themselves. The objective is to write an understandable trade contract to avoid disputes, but procedures should be specified in the event disputes do arise.

The following clause is incorporated in our contract with the electrical contractor for the Smith residence to clarify procedures to be followed in the event of a dispute:

> Any dispute between the Subcontractor and **[Builder]** arising out of or related to this subcontract that is not informally resolved shall be settled by arbitration under the procedures contained in the general contact. Any Subcontractor claims for additional compensation or damages due to acts or omissions of Owner shall be submitted to **[Builder]**, who will

submit the claim to the Owner on behalf of Subcontractor. Resolution procedures will be those contained in the general contract.

Verify Trade Contractor's Independent Status

One of the primary reasons for builders to use written trade contracts is to establish the trade contractor as an independent contractor rather than an employee of the builder. Builders who do not use written trade contracts might not be able to prove the independent status of trade contractors and could end up paying insurance premiums, back taxes, and penalties for not properly withholding Social Security and federal income taxes.

To provide proof that the trade contractor is not an employee of the builder in our sample contract, we included the following provision:

The relationship between **[Builder]** and Subcontractor is that of independent contractor. Subcontractor has the status of an employer as defined by the Unemployment Compensation Act of the state in which this contract is to be performed and all similar acts of the national government including all Social Security Acts.

Subcontractor will withhold from its payrolls as required by law or government regulation and shall have full and exclusive liability for the payment of any and all taxes and contributions for unemployment insurance, workers' compensation, and retirement benefits that may be required by federal or state governments.

Require and Coordinate Written Warranties

The trade contract should require the contractor to provide a warranty for all work performed under the terms of the contract. The builder's contract with the homeowner or buyer usually specifies a warranty period for repair or replacement of any defective work. Normal contracting practice requires the builder to provide a warranty for 1 year after the completion of the project unless otherwise noted.

To minimize builder risk, similar warranty requirements should be included in all trade contracts. Trade contractors then are responsible for repairing or replacing any defective work performed by their employees or agents. Trade contractors must understand that their warranties run concurrently with the builder's warranty and that all warranty periods start with the date of closing with the homeowner or the date the homeowner takes possession of an owner-built house.

The following clauses are used in our sample electrical contract to place a warranty responsibility on the trade contractor:

1. Subcontractor warrants its work under this subcontract against all deficiencies and defects in material and/or workmanship. All materials and equipment furnished shall be new and installed in conformance with code requirements. Subcontractor agrees to repair or replace at its expense and pay for any damages resulting from any defective materials or workmanship that appear within one (1) year after the closing date.

2. Subcontractor shall remedy, at its expense, any defects due to faulty materials or workmanship and pay for any damages caused to other work within seven (7) working days after being notified by **[Builder]** or Owner.

3. **[Builder]** retains the right to perform remedial work not completed within the established time frame and charge the cost to Subcontractor.

Specify Procedures for Contract Suspension or Termination

Although termination is a costly option for all parties and should be seen as a last resort after other options are investigated, procedures for early termination or suspension of the trade contract should be defined in the contract. There may be instances where the builder desires to terminate the contract before the entire scope of work has been completed or to suspend work until a major issue is resolved. Procedures for handling such contingencies need to be clearly understood by the trade contractor.

Procedures for early termination of the trade contract also need to be defined to handle situations when the contractor fails to perform as required by the contract. The trade contract should provide for written notice of noncompliance from the builder to be followed by a specified period in which the contractor is to make corrections. Failure to make corrections then can be used to justify early termination.

The following clauses are used in our example of the Smith residence to provide the builder with the necessary protection to cover these contingencies:

1. If Subcontractor fails to carry out the work in accordance with this subcontract and fails within seven (7) days after receiving written notice to make correction, **[Builder]** will issue a second letter of noncompliance and notify Subcontractor of its intent to terminate the subcontract. If Subcontractor fails to make the necessary corrections within three (3) working days after receipt of the second letter, **[Builder]** may terminate the subcontract.

2. **[Builder]** may, without cause, order Subcontractor in writing to suspend, delay, or interrupt the work in whole or in part for a period of time. An adjustment in subcontract price shall be made to cover the cost of performance, including profit on the increased cost, caused by the suspension, delay, or interruption.

Look at the Entire Contract

Figure 2-1 shows the complete contract document for the electrical work in the Smith residence. Cascade Homes is the name of a fictional builder that has the contract for the project construction. This sample contract was written as a subcontract to the general contract for the construction of the house. If there was not a general contract and the house was being constructed on speculation, the term subcontractor would be replaced with contractor and all references to the general contract would be removed. All appropriate plans and specifications would be referenced in the trade contract.

Cascade Homes

1650 Happy Valley Road
Olympia, Washington 98507

CONSTRUCTION CONTRACT

THIS AGREEMENT, made and entered into this 15th day of June, 19__, by and between CASCADE HOMES and MOUNTAIN ELECTRICAL CONTRACTORS, herein called the "Subcontractor," for all electrical work associated with the construction of the Smith residence located at 7654 Mountain View Road, Tacoma, Washington 98411.

WHEREAS, CASCADE HOMES entered into a contract dated the 3rd day of June 19__ with Thomas J. Smith, herein called the "Owner," for the construction of a residence according to the terms and conditions of said contract and the general specifications and supplements, addenda, general and special conditions, plans, drawings, and other documents made a part thereof, and any change orders or amendments, collectively referred to as the "general contract."

WHEREAS, Subcontractor acknowledges that it is familiar with the general contract and agrees that the general contract is a part hereof and incorporated as a part of this subcontract.

IT IS HEREBY AGREED AS FOLLOWS:

WORK TO BE PERFORMED. Subcontractor agrees to provide all materials, labor, tools, equipment, supervision, supplies, and other items necessary or required to execute the work and perform all electrical work in accordance with the contract drawings, specifications, and any addenda contained in the general contract documents for the construction of the Smith residence. The intent is to provide a complete electrical system that meets local code requirements including, but not limited to, provision and installation of the following:

1. Primary and secondary conductors, conduits, and grounding

2. All panels, subpanels, breakers, and switches for each system including circuits for heating, air conditioning, appliances, water heater, lighting, GFCI, receptacles, switches, and accessories

3. Light fixtures, fans, range hood, smoke detectors, bulbs, and lamps

4. Color coding and marking, as required

5. Cutting, coring, and patching, as required

6. Testing and hot-checking of entire electrical system

Continued

Figure 2-1. Sample Trade Contract Document

Sample Trade Contract, Page 2

CONTRACT PRICE. CASCADE HOMES agrees to pay and Subcontractor agrees to accept as full compensation for performing all work according to the requirements of the contract and furnishing all materials, supplies, and equipment required to execute the work, the sum of *seventeen thousand, six hundred fifty dollars* ($17,650) subject to additions and deductions from such changes in the scope of work as agreed upon in writing through change orders to the contract.

TERMS AND CONDITIONS OF THIS CONTRACT ARE ATTACHED.

Builder: CASCADE HOMES

By: ➤_____
                          Authorized Signature/Title

Date: _____/_____/_____

Subcontractor: MOUNTAIN ELECTRICAL CONTRACTORS

By: ➤_____
                          Authorized Signature/Title

Date: _____/_____/_____

Continued

Figure 2-1. Sample Trade Contract Document *(continued)*

TERMS AND CONDITIONS OF CONSTRUCTION CONTRACT

Between

CASCADE HOMES and MOUNTAIN ELECTRICAL CONTRACTORS

Dated June 15, 19___

1. GENERAL RESPONSIBILITIES

   a. Subcontractor shall comply with the General Conditions of the general contract for the construction of the Smith residence and any interpretations as to the meaning thereof issued by the Owner.
   b. Subcontractor shall ensure that all wiring and electrical items fully comply with code requirements that are effective on the date contract is signed. Code modifications made after contract has been signed are not included.
   c. Subcontractor shall visit the construction site before starting work to understand access restrictions to the site.
   d. Subcontractor shall coordinate the installation of all electrical work with other interfacing trades to ensure an unencumbered worksite.
   e. Subcontractor shall coordinate connection with the local power utility.
   f. Subcontractor shall coordinate with the local cable and telephone utilities to ensure working television and telephone outlets as located on the plans.
   g. Subcontractor shall provide a qualified onsite supervisor whenever work is being performed. Supervisor will attend weekly coordination meetings to review scheduled versus actual progress, quality control, coordination of work, and jobsite safety.
   h. Subcontractor shall submit required shop drawings and manufacturers' cut sheets for all equipment and await receipt of approval of Owner or Architect prior to starting work.
   i. Subcontractor shall at all times keep the project site clean of dirt, debris, trash, and any waste materials arising from the performance of the subcontract. Subcontractor is responsible for removal of all debris created as a result of the work being performed and disposal at a site designated by CASCADE HOMES.

2. COMMENCEMENT AND PROGRESS OF WORK

   a. Subcontractor agrees to comply with and perform the subcontracted work in conformance with project plans and specifications and applicable state, county, and municipal codes, ordinances, and statutes according to the requirements of the CASCADE HOMES' construction schedule or schedules as CASCADE HOMES may from time to time develop and submit to Subcontractor. Within three (3) calendar days after being notified by CASCADE HOMES, Subcontractor shall commence actual construction on such parts of the scope of work as CASCADE HOMES may designate and to thereafter continue diligently in the performance of the work. All work shall be performed in full cooperation with CASCADE HOMES and other subcontractors.
   b. Upon request, Subcontractor shall prepare and submit to CASCADE HOMES for approval a progress schedule to meet the dates as shown by CASCADE HOMES' then-current construction schedule and showing the order in which Subcontractor proposes to carry on the work and the dates on which it will start and complete salient features of the subcontracted work.

Continued

Figure 2-1. Sample Trade Contract Document *(continued)*

c. If in the opinion of CASCADE HOMES Subcontractor falls behind the progress schedule, Subcontractor shall take steps as may be necessary to improve the subcontract progress, and CASCADE HOMES may require Subcontractor to increase the number of shifts and/or overtime work, days of work, and/or increase equipment and/or tools being used, and to submit such revised schedule demonstrating the manner in which the agreed rate of progress will be regained.

3. PAYMENT PROCEDURES

a. Payment will be made to Subcontractor for work performed under this subcontract as measured and certified as being completed by CASCADE HOMES. Before any payment is due, Subcontractor must provide CASCADE HOMES with a written estimate of the total amount of work completed and a signed statement that all completed work fully conforms to contract requirements.

b. Upon application, partial payments for work performed under this subcontract will be made by CASCADE HOMES, and will equal the value of the work performed by Subcontractor, less the sum of previous payments.

c. Subcontractor will receive partial payments fifteen (15) days after invoice has been received by CASCADE HOMES.

d. Upon completion of all contracted work, Subcontractor will be paid the remaining amount due Subcontractor under this subcontract. Final payment shall release CASCADE HOMES from any further obligations whatsoever in respect to this subcontract. Subcontractor shall, as a condition to receipt of final payment, provide to CASCADE HOMES a full release from any and all claims, liens, and demands whatsoever for all matters growing out of, or in any matter connected with this subcontract.

4. LAWS AND REGULATIONS

Subcontractor, its employees and representatives, shall at all times comply with all applicable laws, ordinances, statutes, rules, and regulations, federal and state, county and municipal, and particularly those relating to wages, hours, fair employment practices, nondiscrimination, and working conditions. Subcontractor shall procure and pay for all permits, licenses, and inspections required by any governmental authority for any part of the work under this subcontract and shall furnish any bonds, security, or deposits required by such authority to permit performance of the work.

5. INSURANCE

a. Subcontractor, at its own expense, shall procure, carry, and maintain on all of its operations workers' compensation and employer's liability insurance covering all of its employees, public liability and property damage insurance, and automotive public liability and property damage insurance. Coverage limits shall be in accordance with the requirements of the general contract. Subcontractor is required to name CASCADE HOMES and Owner as additional insureds on Subcontractor's general liability policy.

b. Subcontractor shall provide to CASCADE HOMES prior to commencement of work a certificate from the insurance companies that such insurance is in force and will not be canceled without thirty (30) days written notice to CASCADE HOMES.

Continued

Figure 2-1. Sample Trade Contract Document *(continued)*

6. INDEMNIFICATION

Subcontractor shall indemnify and hold harmless CASCADE HOMES and Owner against any claims, damages, losses, and expenses, including legal fees, arising out of or resulting from performance of subcontracted work to the extent caused in whole or in part by the Subcontractor or anyone directly or indirectly employed by the Subcontractor.

7. SAFETY

    a. Subcontractor and all of its employees shall follow all applicable safety and health laws and requirements pertaining to its work and the conduct thereof, but not limited to, compliance with all applicable laws, ordinances, rules, regulations, and orders issued by a public authority, whether federal, state, or local, including the Federal Occupational Safety and Health Administration, and any safety measures required by CASCADE HOMES or Owner.

    b. Safety of Subcontractor's employees, whether or not in common work areas, is the responsibility of Subcontractor.

    c. Subcontractor agrees to instruct all its employees to inform CASCADE HOMES immediately of any unsafe condition or practice whether or not in common work areas.

8. INSPECTION AND ACCEPTANCE OF COMPLETED WORK

The materials and work shall at all times be subject to inspection by Owner and CASCADE HOMES. Owner and CASCADE HOMES shall be afforded full and free access to the construction site for the purpose of inspection and to determine the general progress of the work. In the event that any of the work or materials are found to be improper or defective by Owner or CASCADE HOMES, Subcontractor shall upon notification in writing from CASCADE HOMES proceed to replace or correct the defective material or workmanship at its own cost and expense. If Subcontractor fails to correct the defective work, CASCADE HOMES, at its option, may replace and correct the same and deduct the cost of correcting the defective work from Subcontractor's payments.

9. CHANGE ORDER PROCEDURES

CASCADE HOMES may order additional work, and Subcontractor will perform such changes in the work as directed in writing. Any change or adjustment to the subcontract price as a result of changes in the scope of work shall be as specifically stated in the change order. If Subcontractor encounters conditions it considers different from those described in subcontract documents or plans, it is required to issue written notice to CASCADE HOMES before proceeding. Subcontractor's failure to issue notice shall constitute waiver of any claims for additional compensation. Only CASCADE HOMES' project superintendent is authorized to issue change orders to Subcontractor. If Subcontractor and CASCADE HOMES cannot agree upon a price for the changes in the work, CASCADE HOMES may direct Subcontractor to execute the changes, and Subcontractor will be paid based on the actual cost to Subcontractor plus a reasonable markup for profit and overhead expenses.

Continued

Figure 2-1. Sample Trade Contract Document *(continued)*

10. DISPUTE RESOLUTION

Any dispute between the Subcontractor and CASCADE HOMES arising out of or related to this subcontract that is not informally resolved shall be settled by arbitration under the procedures contained in the general contact. Any Subcontractor claims for additional compensation or damages due to acts or omissions of Owner shall be submitted to CASCADE HOMES, who will submit the claim to the Owner on behalf of Subcontractor. Resolution procedures will be those contained in the general contract.

11. TRADE CONTRACTOR'S STATUS AS INDEPENDENT CONTRACTOR

The relationship between CASCADE HOMES and Subcontractor is that of independent contractor. Subcontractor has the status of an employer as defined by the Unemployment Compensation Act of the state in which this contract is to be performed and all similar acts of the national government including all Social Security Acts.

Subcontractor will withhold from its payrolls as required by law or government regulation and shall have full and exclusive liability for the payment of any and all taxes and contributions for unemployment insurance, workers' compensation, and retirement benefits that may be required by federal or state governments.

12. WARRANTY

a. Subcontractor warrants its work under this subcontract against all deficiencies and defects in material and/or workmanship. All materials and equipment furnished shall be new and installed in conformance with code requirements. Subcontractor agrees to repair or replace at its expense and pay for any damages resulting from any defective materials or workmanship that appear within one (1) year after the closing date.

b. Subcontractor shall remedy, at its expense, any defects due to faulty materials or workmanship and pay for any damages caused to other work within seven (7) working days after being notified by CASCADE HOMES or Owner.

c. CASCADE HOMES retains the right to perform remedial work not completed within the established time frame and charge the cost to Subcontractor.

13. TERMINATION OR SUSPENSION

a. If Subcontractor fails to carry out the work in accordance with this subcontract and fails within seven (7) days after receiving written notice to make correction, CASCADE HOMES will issue a second letter of noncompliance and notify Subcontractor of its intent to terminate the subcontract. If Subcontractor fails to make the necessary corrections within three (3) working days after receipt of the second letter, CASCADE HOMES may terminate the subcontract.

b. CASCADE HOMES may, without cause, order Subcontractor in writing to suspend, delay, or interrupt the work in whole or in part for a period of time. An adjustment in subcontract price shall be made to cover the cost of performance, including profit on the increased cost, caused by the suspension, delay, or interruption.

Figure 2-1. Sample Trade Contract Document *(continued)*

Chapter 3

Defining the Scope of Work

As discussed in previous chapters, clearly identifying the scope of work is essential when writing trade contracts. In this chapter, the focus is on the development of *scopes of work*, including methods of defining, writing, and organizing. Sample scopes of work for several trades are included at the end of this chapter and on the diskette.

Why Are Scopes of Work Needed?

Builders generally have a good understanding of the trades available in their market area and the scopes of work each trade performs. Several trades might perform the same function, however, so you cannot assume that the trade contractors automatically knows what you expect of them.

If you have a long-standing working relationship with a trade contractor, your past experiences can set a standard for future work, but you should still define the scope of work for each job. If you are building a new relationship with a contractor, no understanding exists and you must define the scope of work.

Having trades perform portions of your building contract transfers work responsibilities to the trade but not the ultimate responsibility for a quality project. Contracting with the trades adds to your risk because you must supervise an independent contractor to ensure that the work is done according to specifications. If the trade does not perform the work as expected, some possible consequences for you are:

- Delays in completing the contract
- Increases in construction and overhead costs
- Reductions in your margin or profit from the job
- Misunderstandings and legal disputes
- Reduced job quality
- Negative reflections on your reputation

The building industry also suffers when work is performed poorly, so it is imperative that the builder and trades have a clear understanding before the work starts. The best insurance that this understanding exists is a written scope of work. Adding a scope of work to trade contracts forces the builder and the trade contractor to think through the work and the building process to attain a better understanding before work starts.

Define the Scope of Work

Three general methods are used to define a scope of work.

Reference Specifications

A common way for builders to define a scope of work for a trade contract is to reference specifications in the general contract. It is a "boilerplate" method that refers the trade to another document to determine the scope of work. An example might be: "Perform all work in section 3 of the specifications." Since the specifications are designed to identify performance standards, products, and manufacturers' specifications rather than define trade scopes of work, the reference to specifications method is not the best way to define trade scopes of work. Some of the disadvantages to this method are:

- Specifications often are written by an outside party who is not familiar with the builder's business and potential trade contractors.
- This method is dependent on the quality of the specifications and how well they are written.

Reference Parts of the House

Instead of referring to specifications, referencing parts of the house more precisely describes what is expected of a trade, but lacks the details needed for a precise understanding. It also leaves too many options open for interpretation since the trade must define the work. For example: "Perform all work related to the foundation."

Provide a Detailed List of Work Items

We recommend that builders who want the clearest scopes of work provide a detailed list of work items. This method describes the work and is easily understood. The scope can be short or lengthy depending on the complexity of the work. It is the most descriptive and can be applied to any set of trades. For example: "Layout, excavate, form, and pour all footings. Strip all footing forms within 24 hours and clean up around the work area."

Write the Scope

There are many ways to write scopes of work for the trades. Small, simple houses usually need short (less than one page) scopes of work. Large, multiphase projects, on the other hand, need more in-depth scopes of work. The content of a scope of work comes from:

- Contract documents
- Building codes
- Trade skills and resources
- Builder preferences

Follow Related Contract Documents

Reviewing the drawings and specifications is a good place to start. Find out what the documents require. The scope of work is structured somewhat differently for a custom home and one built on speculation. If the builder has a contract with a homeowner, the contract documents (specifications, drawings, general conditions, and agreement) are used to define specific scope of work issues. This holds true whether the builder, homeowner, or architect wrote the contract documents.

For a speculation house, however, the builder is the owner and therefore establishes the scope of work. In both situations, the builder constructs the house in accordance with the scope of work as defined by the contract documents.

Incorporate Necessary Information from Building Codes

Check the Americans with Disabilities Act (ADA), safety regulations, and building codes. Although the scope of work is typically thought to be defined by the drawings and specifications, some parts are actually defined by codes. Designers, builders, and trade contractors must consider codes when defining scopes of work. The type, location, and installation of many building components are controlled by codes. Smoke detectors, railings, and devices for structural connectors are examples of scope of work items regulated by codes.

When codes influence the way a trade's work is performed, it is good practice to make a reference to that code when you write the scope of work. This helps define the work and places contractual responsibility on the trade contractor to follow code requirements.

Understand Trade Skills and Resources

Before you write a scope of work for the first time, get trade contractors involved. Call a meeting and ask for their input. Discuss their labor resources and skill levels for the job. Ask about the availability of equipment and other resources needed to perform the tasks in the scope of work. Be sure to evaluate optional construction methods for performing the work.

Specialty trades usually know more about their business than anyone else and can make suggestions that best fit their ability and willingness to perform the work. Make the trade part of your team and you can tailor the scope's conditions to fit the parties and the project. This makes the scope of work a more useful tool for a successful project.

It is a wise management practice to select trades who have a good reputation for quality work and the skills to get the job done on time. Also, it's a good idea to review the details of all written scopes of work with an experienced project superintendent.

Include Builder Preferences

Once a scope of work document has been drafted and checked by the trade, the builder still must make adjustments. Knowing which scope of work items "belong" to which trade is not always obvious. Deciding which trade will perform a specific activity ultimately depends on how the builder wants the work performed. For example: Who will install the hot water tank for a radiant heating system? Will you use the heating, ventilating, and air conditioning (HVAC) contractor or the plumbing contractor? Who will install the caulking between the tub or water closet and the floor tile? Will you use the painter, plumber, or tile contractor? These decisions are properly made by the builder and reflected in the detailed scopes of work developed for these trade contractors.

Since many scope of work items are commonly performed by two or more trades, each scope of work needs to be written carefully to assure that a scope of work item is not placed in two or more trade contracts. For example, if both the framing and trim carpenter include a price to hang the front door, the builder pays twice. Likewise, all scope of work items need to be in some trade's scope of work. If an item is missing from a trade's scope, it eventually will become a change order and paid out of the builder's profit.

Organize the Scope of Work

The scope of work includes:

- A general introduction in broad terms
- A list of specific scope items to be included
- Quality issues
- A list of items that are to be excluded from the price or not performed by the trade

Introduce the Scope of Work

Specify whether the work is labor only, material only, or labor and material in the scope's introduction. Include all general types of cost issues including equipment, supervision, tools, etc. The introduction does not identify the details. If this section is well written, it describes what a trade will do in general, but lacks the details. The purpose of the introduction is to:

- Identify the trade performing the work (not the firm's name)
- Define the scope of work in broad terms
- Make reference to the contract documents

List Detailed Scope of Work Items

After the introduction, write a detailed list of scope of work items in a logical sequence. Whenever possible, the list should follow the expected construction sequence. Basically the listed items should be very specific and include all the work being asked of the trade. This list is the essence of the scope of work.

Use action words to describe scope of work items. For items that relate to actions or tasks, use action words that are specific and cannot be misinterpreted. "Install," "hang," and "erect" all imply labor-only activities. "Provide" and "supply" imply the material will be paid for by the installer. Take care to include "provide" and "install" when both labor and material are intended. Do not allow scopes of work to be ambiguous.

Many of the following sample scopes of work have optional items that might be useful for some light commercial or multifamily projects. Do not include these optional scope items in residential trade scopes unless they apply.

Address Quality Issues

Some builders have standards or specifications that are more restrictive than those specified by the contract documents. When a minimum standard needs to be included in a scope of work, add it to the list of detailed items or include it as part of a special quality section within the scope. Minimum quality standards in scopes of work also can be used to define the builder's warranty standards. If included in the scope, the trade must also meet these standards. Some examples of the minimum quality standards can be found in the *Residential Construction Performance Guidelines* published by the National Association of Home Builders. Examples also are included in some of the following sample scopes of work included in this chapter.

Incorporate Exclusions in the Scope of Work

The exclusion section in scopes of work should highlight items that builders do not want included in the scope of work. You should list all the items that might be misunderstood. Do not assume the trade contractor knows what you excluded. If excluded items are not clearly listed, trade contractors may include them in their price quotes "just in case" the builder expects them.

Develop Scopes of Work

Use the following sample scopes of work as guides when you write your request for quotes and trade contracts. Some of the sample scopes of work provided here do not apply to typical residential projects but might apply to special high-end residential, multifamily, or light commercial projects. They have been presented as generic lists to fit the needs of most builders in most geographical locations of the United States. You will need to modify the sample scopes to reflect your preferences, local variations in climate and in building codes, differences in local trade practices, and differences in the skills available from trades in your area.

The following sample scopes of work are presented in this chapter:

- Site Preparation and Clearing
- Site Excavation
- Site Development
- Site Landscape
- Site Hardscape
- Concrete Flat Work
- Concrete Foundations
- Masonry—Brick
- Masonry—Concrete Masonry Units
- Metal—Railing
- Framing—Light-Gauge Metal
- Framing—Rough Carpentry
- Carpentry—Siding
- Carpentry—Finish
- Insulation—Batt/Blown
- Roofing—Shingle
- Roofing—Built-up
- Glazing
- Finishes—Drywall
- Finishes—Paint
- Finishes—Ceramic Tile
- Finishes—Cementitious Stucco
- Finishes—Synthetic Stucco (Exterior Insulated Finish System)
- Finishes—Custom Millwork Wood Stairs
- Finishes—Cabinetry
- Finishes—Interior Cleanup
- Flooring—Carpet, Vinyl, and Wood
- Appliances
- Prefabricated Metal Fireplace
- Heating, Ventilating, and Air Conditioning
- Fire Protection
- Plumbing
- Electrical

Site Preparation and Clearing

Scope:

1. Provide all labor, material, hand tools, equipment (including safety equipment), supervision, supplies, and other incidentals necessary to install the site preparation and clearing work and perform all such work in accordance with the contract drawings, specifications, addenda, and other documents that make up the contract documents but not limited to the following:
   a. Install tree protection prior to any clearing.
   b. Strip site of all ground vegetation as directed by project superintendent.
   c. Remove topsoil to the specified depths and stockpile on site in an area designated by the project superintendent for future use.
   d. Remove all brush, vegetation, and trees within the construction limits.
   e. Remove all stumps and root systems.
   f. Burn brush, vegetation, and tree debris on site in an area as directed by the project superintendent and according to local fire department requirements. Attain required permits.
   g. Rake all stripped and scarified areas prior to grading.
   h. Remove any unburned material from construction site.
2. Save all trees and brush, except those designated.

3. Take special care not to damage any trees designated to be saved by the project superintendent.
4. Provide temporary service and access roads as required for site clearing, grading, and storm drainage.

Optional Scope Items:

5. Provide any surveying necessary to bring the site to the specified grades.
6. Protect the benchmarks and property lines set by the builder.
7. Provide the demolition of all obstructions as shown and remove them from the site.

Exclusions:

None

Site Excavation

Scope:

1. Provide all labor, material, hand tools, equipment (including safety equipment), supervision, supplies, and other incidentals necessary to install the excavation work and perform all such work in accordance with the contract drawings, specifications, addenda, and other documents that make up the contract documents but not limited to the following:
   a. Cut all excavated areas as specified to meet the foundation elevations.
   b. Over excavate beyond footing to allow working clearances needed to install foundations and walls.
   c. Provide means to retain unexcavated soil, slope wall, or bench step the excavation to prevent soil caveins.
   d. Use clean fill for all backfill.
   e. Backfill soils as specified to attain rough grade.
   f. Attain approval from project superintendent or building department before backfilling foundation walls.
   g. Place and spread fill dirt to attain proper grade as shown on the site drawings and specifications.
   h. Compact soils as required in the specifications.
   i. Install French drains as specified.
2. Limit fill depth in foundation pads to no more than 6-inch lifts.
3. Maintain structural fill in the building area to plus or minus 1/10 of a foot of the specified elevation.
4. Rough grade site, including swales, before structure is started.
5. Provide and maintain all erosion control (rip rap) required by governmental authorities having jurisdiction and as shown on the drawings, including grassing.

6. Take special care not to damage any trees designated to be saved by the project superintendent.

Optional Scope Items:

7. Provide any surveying necessary to bring the site to the specified grades.
8. Protect the benchmarks and property lines set by the builder.
9. Protect all adjacent property from damage resulting from this work.
10. Include a minimum of three compaction tests for each building pad.

Exclusions:

1. Fine grading (included in Site Landscape)
2. Excavation and backfill of utility trenches (included in Site Development)
3. Utility tap fees and impact fees (by builder)

Site Development

Scope:

1. Provide all labor, material, hand tools, equipment (including safety equipment), supervision, supplies, and other incidentals necessary to install the site development work and perform all such work in accordance with the contract drawings, specifications, addenda, and other documents that make up the contract documents but not limited to the following:
   a. A complete storm drainage system including all required pipe, tees, manholes, manhole covers, catch basins, grates, outflows, and connections
   b. A complete sanitary sewer system including all required sanitary pipe, manholes, valves, pumps, and connections
   c. A complete water system including all required pipe valves, pumps, meters, and connections
   d. A complete storm water system including piping from a point 10 feet outside the building to the point of discharge, including flared ends, manholes, catch basins, head walls, and culverts
   e. Site electric including site lighting, power to equipment (including safety equipment), and underground wiring to the building
   f. A complete gas system including piping, tanks, and valves
   g. All dewatering required to perform this scope
   h. Excavation and backfill all utility trenches needed for this scope
   i. Concrete transformer pad
   j. Flushing, purging, chlorinating, and preparing the drainage, sanitary, and water systems as need to make the systems fully operational
   k. Tests required by codes and specifications
   l. Temporary water to site as needed
   m. Site permits related to work

2. Work within project schedule, which might require phasing of paving work.
3. Coordinate the installation of all work with other interfacing trades and specifically coordinate with the local utility companies (sewer, water, and gas) relative to installation of their lines.
4. Coordinate with related regulatory bodies to install all work in accordance with those bodies and receive final tests and approvals in writing for sewer, water, drainage, compacting, and excavation.
5. Obtain and turn over to the builder all required governmental approvals.
6. Do not cover up any part of systems until required approvals are received.

Exclusions:

1. Rough grades will be established by others and shall be to elevations as specified on the drawings to plus or minus 1/100 of a foot of the grades needed to finish.
2. Site clearing and site excavation (included in Site Preparation and Clearing).
3. All rip rap as shown on the drawings (included in Site Excavation).
4. Demolition.
5. Site paving (included in Site Hardscape).
6. Site preparation (included in Site Preparation and Clearing).

Site Landscape

Scope:

1. Provide all labor, material, hand tools, equipment (including safety equipment), supervision, supplies, and other incidentals necessary to install the landscaping work and all such work in accordance with the contract drawings, specifications, addenda, and other documents that make up the contract documents but not limited to the following:
   a. Sodding, seeding, and hydro-seeding
   b. Plants, shrubs, and landscape materials
   c. Trees
   d. Ground cover
   e. Edging
   f. Decorative stones
   g. Railroad ties
   h. Topsoil
   i. Fertilizer
   j. Bark, chips, and mulching
   k. Erosion control
2. Hand grade and adjust grade for the thickness of the sod so that the sod becomes the final grade as indicated on the drawings. The sod

shall come up to the elevation of adjoining sidewalks unless the sidewalk blocks the planned flow of water. In such cases, notify the project superintendent.

3. Prepare site for planting materials.
4. Rake and remove all debris, rocks, and foreign matter before laying any sod or seed.
5. Adjust the finish grade contours of the site to allow proper runoff of water.
6. Allow a minimum of 6 inches from floor slab to final grade.
7. Roll sod within 24 hours after initial laying to achieve a smooth sod blanket.
8. Ensure that sod is free from foreign material and weeds.
9. Lay sod neat and tight.
10. Provide final grade and swells to carry water away from buildings. If any obstruction prevents the proper grading of the site, immediately notify the project superintendent.

Optional Scope Items:

11. Water sod prior to rolling as needed. The work is the responsibility of trade contractor until homeowner's acceptance is received.
12. Install irrigation system to include:
    a. Connection to water supply
    b. Pipe, heads, valves, and time clock
    c. Final electrical hookup
    d. Excavation and backfill for irrigation system
    e. Sleeves under flatwork

Exclusions:

1. Backfill
2. Compaction

Site Hardscape

Scope:

1. Provide all labor, hand tools, equipment (including safety equipment), supervision, supplies, and other incidentals necessary to install the site hardscape work and perform all such work in accordance with the contract drawings, specifications, addenda, and other documents that make up the contract documents. The intent of this scope is to complete site improvements and not limited to the following:
    a. Sidewalks, walkways, curbs, retaining walls, planters, patio slabs, and steps
    b. Preparation, placement, and finishing concrete flatwork
    c. Stripping on paved areas as specified
    d. Paving bricks, stepping stones, decorative stone

  e. Limerock, gravel, and soil cement subbase
  f. Fine grade, compaction, and fill under paved areas
  g. Layout
  h. Establishing grades
  i. Expansion joints
  j. Removal of all excess concrete resulting from the placement
  k. Installation of sidewalks to proper elevations to match required final grades, especially in swales, to allow the proper flow of water off sidewalks
  l. Placement and finish of all concrete sidewalks according to the drawings and specifications
 2. Formwork to include the following:
  a. Forms needed to perform work
  b. Tractor work needed to install forms
  c. Stripping and removal of all forms within 48 hours after each placement

Optional Scope Items:

 3. Turn all transit mix delivery tickets in daily to the project superintendent.
 4. Do not add more than 1 gallon of water for each cubic yard of concrete.

Exclusions:

 1. Materials paid for by others
 2. Concrete structural or flatwork in building
 3. Railroad ties

Concrete Flat Work

Scope:

 1. Provide all labor, hand tools, equipment (including safety equipment), supervision, supplies, and other incidentals necessary to install the concrete flat work and perform all such work in accordance with the contract drawings, specifications, addenda, and other documents that make up the contract documents but not limited to the following:
  a. Poured-in-place slab on grade, slab on deck, stairs, pan stair treads, and mechanical pads
  b. Construction joints, expansion joints, and premolded joint filler
  c. Rebar, chairs, mesh, inserts, anchors, dowels, and connecting devices embedded in the flat concrete work
  d. Corner reinforcing bars (stand-up steel)
  e. Curing and hardening compound
  f. Concrete cutting
  g. Unloading, setting, and aligning of all embedded plates

      h. Bulkheads and screeds
      i. Perimeter insulation
      j. Vapor barrier under slabs
      k. Gravel under slabs
      l. Layout work
      m. Establishing grades

2. Install bolts prior to placing concrete. Do not push bolts in after concrete is placed.
3. Apply the specified broom, steel trowel, and exposed surface finishes.
4. Fine (hand) grade to assure proper thickness of concrete.
5. Material overruns are the expense of trade contractor.
6. Clean any exposed concrete surfaces.
7. Remove all excess concrete resulting from the placement.
8. Protect all adjacent work.
9. Notify ready-mix company when concrete is to be delivered.
10. Provide formwork:
    a. Excavate any soils needed to install forms and place concrete in this work.
    b. Prepare forms to ensure proper quantity of concrete is used.
    c. Erect and remove all formwork.
    d. Rub and smooth all exposed vertical concrete after stripping.
    e. Tool all exposed edges.

Optional Scope Items:

11. Maintain a uniform level (1/8 inch in 8 feet) using a steel trowel finish for all slab areas to receive carpet, vinyl, and tile.
12. Turn in all transit mix delivery tickets to the project superintendent daily.
13. Do not add more than 1 gallon of water to each cubic yard of concrete.
14. Install control joints with 16-by-20-inch spacing.

Exclusions:

1. Building corner offset and an elevation benchmark provided by others
2. Materials paid for by others
3. Conveying and pumping equipment
4. Site hardscape concrete

Concrete Foundations

Scope:

1. Provide all labor, hand tools, equipment (including safety equipment), supervision, supplies, and other incidentals necessary to install the concrete foundation systems and perform all such work in accordance with the contract drawings, specifications, addenda,

and other documents that make up the contract documents but not limited to the following:

  a. Pad, isolated, continuous, strip, and spread footings
  b. Piers and foundation walls
  c. Rebar, inserts, anchors, and connecting devices in the concrete associated with this work
  d. Control joints, construction joints, and expansion joints in the work
  e. Bulkheads and screeds
  f. Waterstops
  g. Unloading, setting, and aligning of all embedded plates
  h. Layout work
  i. Establishing grades
  j. Blockouts in walls

2. Square building and check all dimensions. Do not measure just the sides and diagonals.
3. Dispose of excess spoil material in an area designated by the project superintendent.
4. Level top of footing within 1/2 inch.
5. Rub and smooth all exposed vertical concrete and fill voids.
6. Remove and clean up excess concrete resulting from the placement.
7. Install steps in the foundation as needed to follow the slope of the lot.
8. Place all concrete on undisturbed soil.
9. Material overruns are the expense of trade contractor.
10. Waterproof wall.
11. Provide formwork:

  a. Excavate any soils needed to install forms and place concrete in this work.
  b. Prepare forms to ensure proper quantity of concrete is used.
  c. Erect and remove all formwork.
  d. Rub and smooth all exposed vertical concrete after stripping.
  e. Tool finish all exposed edges.
  f. Remove all snap tie ends.

Optional Scope Items:

12. Turn in all transit mix delivery tickets to the project superintendent daily.
13. Do not add more than 1 gallon of water to each cubic yard of concrete.
14. Protect all adjacent work.

Exclusions:

1. Building corner offset and elevation benchmark provided by others
2. Concrete flat work

    3. Site hardscape
    4. Materials paid for by others

Masonry—Brick _____

Scope:

1. Provide all labor, hand tools, equipment (including safety equipment), supervision, supplies, and other incidentals necessary to install the brick masonry work and perform all such work in accordance with the contract drawings, specifications, addenda, and other documents that make up the contract documents but not limited to the following:
   a. Exterior brick work and necessary incidentals for a complete job
   b. Steel angle headers
   c. Weepholes
   d. Mortar, colored mortar, and additives
   e. Poured-in-place bond beams including formwork
   f. Grout in cavity walls and columns
   g. Anchors, bolts, ties, inserts, flashing, and connecting devices that are in the work
   h. Expansion joints
   i. Water repellent coatings on brick masonry
   j. Brick ties to backup wall
2. Cut all brick as needed.
3. Provide all necessary scaffolding, planks, and material handling equipment.
4. Provide all material handling equipment needed for work.
5. Coordinate the storage of materials with project superintendent.
6. Brace columns.
7. Clean all brick and floors of excess mortar and grout with an approved acid wash.
8. Stack all unused brick on pallets.

Optional Scope Items:

9. Parge masonry walls.

Exclusions:

1. Concrete masonry units (CMU)
2. Materials paid for by others

Masonry—Concrete Masonry Units _____

Scope:

1. Provide all labor, hand tools, equipment (including safety equipment), supervision, supplies, and other incidentals necessary to install the concrete masonry unit (CMU) work and perform all such work in accordance with the contract drawings, specifications, ad-

denda, and other documents that make up the contract documents but not limited to the following:

a. CMU walls, columns, and pilasters
b. Concrete block, precast lentils, headers, sills, and related concrete units needed for a complete system
c. Poured-in-place bond beams
d. Mortar, colored mortar, and additives
e. Water repellent coatings on CMU
f. Waterproofing and dampproofing coatings on CMU
g. Embedded items including flashing, expansion joints, ties, bolts, anchors, and necessary incidentals in masonry or concrete
h. Grouting concrete and embedded reinforcing steel for lintels, headers, tie beams, pilasters, columns, and reinforced filled cells
i. Loose fill insulation placed in the CMU
j. Bracing to prevent wall from tipping or falling over

2. Provide all necessary scaffolding and planks.
3. Clean all excess mortar and concrete spillage from all masonry and floors.
4. Provide formwork:
   a. Excavate any soils needed to install forms and place concrete in this work.
   b. Erect and remove all formwork.
   c. Prepare forms to ensure proper quantity of concrete is used.
   d. Rub and smooth all exposed vertical concrete after stripping.
   e. Tool finish all exposed edges.
5. Stack all unused units on pallets.

Optional Scope Items:

6. Set hollow metal frames in masonry walls (frames supplied by others).
7. Install all metal frames plumb, level, and braced, and prevent from distorting until concrete is sufficiently hardened. Remove spreader bars and braces.

Exclusions:

1. Materials paid for by others
2. Brick work

Metal—Railing

Scope:

1. Provide all labor, material, hand tools, equipment (including safety equipment), supervision, supplies, and other incidentals necessary to install the railing work and perform all such work in accordance with the contract drawings, specifications, addenda, and other docu-

ments that make up the contract documents but not limited to the following:
a. Aluminum railing and accessories required
b. Coring fasteners, bolts, and anchors needed to secure railing and meet codes
2. Coordinate the heights of stairrail and guardrail with the builder.
3. Protect all work from damage until acceptance.

Optional Scope Items:
4. Provide shop drawings that are stamped by a registered engineer.
5. Prepare the surface.

Exclusions:
None

Framing—Light-Gauge Metal

Scope:
1. Provide all labor, material, hand tools, equipment (including safety equipment), supervision, supplies, and other incidentals necessary to install the metal framing work and perform all such work in accordance with the contract drawings, specifications, addenda, and other documents that make up the contract documents but not limited to the following:
a. Metal framing for the interior and exterior portions of the building including walls, roof, soffits, dropped ceiling, floor system, balcony, stairs, furring, and platforms for boilers, hot water heater, and air handlers
b. Metal framed studding, rafters, joists, decking, and trusses
c. Wood blocking for firestops, smokestops, and fire-rated walls and floors required by the local codes
d. Wood blocking necessary to secure all hanging and recessed devices such as cabinets, fixtures, handrails, stairs, and tubs as required by the drywall trade
2. Install all door and window openings for acceptable tolerance in accordance with manufacturer's recommendations
3. Ask project superintendent for any information needed to complete work.
4. Clean up all materials resulting from the work and remove from the site.

Optional Scope Items:
5. Provide all necessary fastening devices such as nails, shots, and pins.

Exclusions:
1. Sheathing
2. Siding

Framing—Rough Carpentry

Scope:

1. Provide all labor, hand tools, equipment (including safety equipment), supervision, supplies, and other incidentals necessary to install the rough carpentry work and perform all such work in accordance with the contract drawings, specifications, addenda, and other documents that make up the contract documents but not limited to the following:
   a. Framing including walls, floors, decking, trusses, furring, dropped ceiling, soffit, balcony, wood stairs, and platforms needed for hot water heater and air handlers
   b. Blocking and firestopping necessary for quality framing and securing all hanging and recessed devices such as cabinets, fixtures, handrails, stairs, and tubs
   c. Blocking as required by the building inspector and the installation of drywall
   d. Sheathing for floor, roof, and walls
   e. Shear walls
   f. Exterior rigid board insulation
   g. Exterior soffit trim and cornice
   h. Dry-in roofing felt
   i. Connectors, straps, anchors, and other devices required for earthquake and hurricane code compliance
2. Frame walls level. If slab is out of level, meet with project superintendent before starting work.
3. Frame door and window openings to acceptable tolerance in accordance with manufacturers' recommendations.
4. Provide any equipment necessary to hoist materials for this work.
5. Provide all necessary fastening devices such as nails, shots, and pins.
6. Use coated or galvanized devices exposed to weather or touching cedar or redwood in accordance with the American Plywood Association or the Western Cedar Association.
7. Coordinate with the drywall contractor the placement of all firestops, smokestops, and fire-rated walls and floors, and to assure all required blocking is installed properly.
8. Coordinate the work sequences with the drywall and mechanical trades.
9. Clean up all materials resulting from the work and remove from the site.

Optional Scope Items:

10. Install all siding and exterior trim for cornice, fascia, soffits, railings, windows, doors, and trim (included in Carpentry—Siding).
11. Install all exterior doors, thresholds, sliding glass doors, windows, and screens.

12. Install all rigid board and insulation sheathing (included in Carpentry—Siding).

Exclusions:

1. Except as noted, materials paid for by others
2. Masonry ties
3. Finish hardware
4. Exterior siding
5. Windows and exterior doors

Carpentry—Siding

Scope:

1. Provide all labor, hand tools, equipment (including safety equipment), supervision, supplies, and other incidentals necessary to install the wood, vinyl, and metal siding work and perform all such work in accordance with the contract drawings, specifications, addenda, and other documents that make up the contract documents but not limited to the following:
    a. Exterior siding
    b. Exterior trim for cornice, fascia, soffits, railings, windows, and doors
    c. Exterior doors, thresholds, sliding glass doors, windows, and screens
    d. Flashing including window, door, and wall openings
    e. Building wrap
    f. Wall felt
2. Caulk all joints and nail holes.
3. Do not cover up inferior or defective work.
4. Comply with the manufacturers' recommended installation procedures.
5. Use coated or nonferrous nails, screws, and other fasteners that are exposed to weather or touch cedar or redwood in accordance with the American Plywood Association or the Western Cedar Association.
6. Clean up all materials resulting from the work and remove from the site.

Optional Scope Items:

7. Install all rigid board and insulation sheathing
8. Install exterior sheathing.

Exclusions:

1. Materials provided by others
2. Roof flashing
3. Masonry flashing

Carpentry—Finish

Scope:

1. Provide all labor, hand tools, equipment (including safety equipment), supervision, supplies, and other incidentals necessary to install the finish carpentry work and perform all such work in accordance with the contract drawings, specifications, addenda, and other documents that make up the contract documents but not limited to the following:
   a. Finish carpentry including base molding, door molding, chairrail, handrail, wood railing, shoe, corner, and attic access casing
   b. Bath accessories including medicine cabinets, towel bars, toilet paper holders, and handicap grab bars
   c. Interior doors
   d. Interior decorative spindles, wood caps, and disappearing stairways
   e. Finish hardware including deadbolts, locksets, latchsets, thresholds, doorstops, closures, templates, push/pull plates, panic hardware, weather stripping, and kick plates
   f. Wood and metal shelving
2. Set and adjust all doors to fit 1 inch off final floor over carpet and 1/2 inch over vinyl and ceramic tile surfaces.
3. Shim all door jams at butts and latch as required for a solid fit.
4. Miter interior corners using the "coped" method.
5. Do not cover up inferior or defective work.
6. Clean up all materials resulting from the work and remove from the site.

Optional Scope Items:

7. Provide and install chalkboards, tackboards, signage, and mail boxes.
8. Provide and install fire extinguishers.
9. Provide and install identification numbers and letters for door and buildings.
10. Turn over all keys marked for appropriate doors to project superintendent.
11. Provide and install exterior metal prehung door units.
12. Provide and install door slabs in metal frames.
13. Set thresholds in a full bed of sealant.
14. Adjust thresholds so that exterior doors fit snugly after cleanup.

Exclusions:

1. Materials paid for by others
2. Ceramic bath accessories embedded in ceramic tile
3. Sliding glass doors

   4. Exterior doors
   5. Cabinets and custom millwork

Insulation—Batt/Blown

Scope:

1. Provide all labor, material, hand tools, equipment (including safety equipment), supervision, supplies, and other incidentals necessary to install the insulation work and perform all such work in accordance with the contract drawings, specifications, addenda, and other documents that make up the contract documents but not limited to the following:
   a. Batt and blown insulation to form a continuous envelope between the interior conditioned air space and the exterior unconditioned air space
   b. A separate vapor barrier if one is not included in the above insulation
   c. Sound insulation as shown on the drawings
2. Install all work according to the specifications, procedures, and practices prescribed by the manufacturer(s).
3. Fill all cracks and joints in excess of 1/4 inch with insulation to maintain the integrity of the exterior envelope.
4. Place baffles at soffits as needed to assure a 1-inch space for air intake.
5. Allow a 1-inch space above all insulation in cathedral ceilings for air flow.
6. Seal with expandable foam all openings around windows, doors, piping, ducts, receptacles in exterior envelope walls, and other areas as required by code.
7. Clean up all materials resulting from the work and remove from the site.

Optional Scope Items:

8. Use safing insulation in all fire-rated walls and floors to seal penetrations for pipe, wire, and duct as required by the building codes or fire department.

Exclusions:

1. Insulation in masonry units (included in Masonry—Concrete Masonry Units)
2. Any rigid wall or roof insulation boards
3. Roof insulation (included in Roofing—Built-up)
4. Rigid insulation (included in Carpentry—Siding)

Roofing—Shingle _____

Scope:

1. Provide all labor, hand tools, equipment (including safety equipment), supervision, supplies, and other incidentals necessary to install the shingle roofing and perform all such work in accordance with the contract drawings, specifications, addenda, and other documents that make up the contract documents but not limited to the following:
   a. Roof shingles
   b. Flashing, counter flashing, and eaves drip
   c. Vents, boots, skylights, all fasteners, accessories, and necessary incidentals to provide a complete roof system
2. Install shingles according to the specifications, procedures, and practices prescribed by the manufacturer.
3. Stack shingles materials on roof.
4. Remove all excess materials and nails from roof and worksite.
5. Do not allow dissimilar metals to touch.
6. Nail in accordance with design and wind load for area codes.

Optional Scope Items:

7. Lap all flashings, drips, and stops a minimum of 6 inches and install bull at each joint.
8. Provide and install gutters and downspout.

Exclusions:

1. Materials paid for by others

Roofing—Built-up _____

Scope:

1. Provide all labor, material, hand tools, equipment (including safety equipment), supervision, supplies, and other incidentals necessary to install the built-up roofing and perform all such work in accordance with the contract drawings, specifications, addenda, and other documents that make up the contract documents but not limited to the following:
   a. Single-ply, built-up roofing, and membrane roofing
   b. Cant strips, expansion joints, roof hatches, skylights, accessories, and necessary incidentals for a complete roof
   c. Roofing insulation
   d. Flashing, counter flashing, reglets, eave drips, gravelstops, vents, and boots
   e. Adhesives, mastic, and solder
   f. Fasteners necessary to complete the work
2. Stack roofing materials on top of roof.

3. Install all roof materials according to the specifications, procedures, and practices prescribed by the manufacturer(s).
4. Install all materials to meet the wind load standards for the area.
5. Do not allow dissimilar metals to touch.
6. Clean up all materials resulting from the work and remove from the site.

Optional Scope Items:

7. Lap all flashings, drips, and stops a minimum of 6 inches and set in bitumen at each joint.
8. Vary insulation thickness to provide drainage to interior drains at the rate of 1/8 inch per foot.
9. Provide and install gutters and downspouts.
10. Provide and install wood blocking and nailers.

Exclusions:

None

Glazing

Scope:

1. Provide all labor, material, hand tools, equipment (including safety equipment), supervision, supplies, and other incidentals necessary to install the glazing work and perform all such work in accordance with the contract drawings, specifications, addenda, and other documents that make up the contract documents but not limited to the following:
   a. Fixed glass, mirrors, and glazing
   b. Seals, gaskets, stops, and accessories as required to complete the work

Exclusions:

1. Windows
2. Sliding glass doors

Finishes—Drywall

Scope:

1. Provide all labor, material, hand tools, equipment (including safety equipment), supervision, supplies, and other incidentals necessary to install the drywall work and perform all such work in accordance with the contract drawings, specifications, addenda, and other documents that make up the contract documents but not limited to the following:
   a. Hanging all drywall (gypsum board/sheetrock) including all walls, ceilings, and soffits
   b. Finishing all drywall including taping, mudding, skimming, sanding, and texturing

    c. Nails, screws, fasteners, and accessories needed for a complete job

    d. Cementitious backer board in all locations specifying ceramic tile

    e. Knockdown, popcorn, stipple, orange peel, or other similar texture in areas as directed by project superintendent

2. Install and finish all work according to the specifications, procedures, and practices prescribed by the manufacturer(s).

3. Coordinate with local building officials and fire marshal to ensure all work is in compliance with local codes including penetrations, dryer vents, attic separation, draftstops, smokestops, and protected (fire-rated) walls and floors.

4. Remove all scrap resulting from the work from jobsite within 24 hours after applying texture.

5. Walk job with project superintendent and framing trade contractor to assure work is ready to start.

6. Do not cover up inferior or defective work.

7. Inspect the work with the project superintendent and painting contractor to identify and correct all defective work prior to applying any texture.

8. Clean up and dispose of all trash, debris, and necessary incidentals from jobsite.

9. Clean mud, tape compound, and textures off floors.

10. Protect tubs from damage as instructed by project superintendent by installing a piece of drywall, protective coating, or liner in each tub prior to working in bath areas.

Optional Scope Items:

11. Install water resistant gypsum board (MR) in all locations specifying ceramic tile.

12. Provide and install marble window sills. (This item may be included with ceramic tile.)

13. Provide and install thermal and acoustic batt insulation.

Exclusions:

1. Exterior wall sheathing with drywall (gypsum board/sheetrock) applied to exterior walls, ceilings, and soffits as outlined on the drawings (This item may be included with Framing—Rough Carpentry.)

2. Light-gauge metal framing (This item may be included with Framing—Light-Gauge Metal or Finishes—Drywall.)

3. Wood blocking (This item may be included with Framing—Rough Carpentry.)

Finishes—Paint

Scope:

1. Provide all labor, material, hand tools, equipment (including safety equipment), supervision, supplies, and other incidentals necessary

to install the painting work and perform all such work in accordance with the contract drawings, specifications, addenda, and other documents that make up the contract documents but not limited to the following:

    a. Prime, paint, seal, and stain finished surfaces according to the finish schedule

    b. Finish all exposed drywall, concrete masonry units, concrete, and materials including all walls, ceilings, and miscellaneous surfaces for a complete job

2. Apply all materials according to the specifications, procedures, and practices prescribed by the manufacturers.
3. Seal the edges of all wood doors in accordance with the door manufacturer's specifications.
4. Caulk all cracks and exposed seams in painted wood trim.
5. Caulk as required around all windows and door joints to maintain a sealed surface.
6. Touch up all work as needed.
7. Remove and clean all over spray, drops, runs, and spillage from hardware, glass, cabinets, doors, and nonpainted surfaces.
8. Check all wood doors during final touchup to assure none were cut after the doors were painted.
9. Do not cover up inferior or defective work.
10. Clean up all materials resulting from the work and remove from the site.

Optional Scope Items:

11. Replace all doors, hardware, mirrors, and accessories that were removed to accomplish the work.
12. Make a color board and receive written approval from builder before starting work.

Exclusions:

None

Finishes—Ceramic Tile

Scope:

1. Provide all labor, material, hand tools, equipment (including safety equipment), supervision, supplies, and other incidentals necessary to install the ceramic tile work and perform all such work in accordance with the contract drawings, specifications, addenda, and other documents that make up the contract documents but not limited to the following:

    a. Marble, granite, ceramic, paver, and quarry tile, and accessory pieces for a complete installation

    b. Tile for all floors, walls, showers, tub areas, fireplace surrounds, hearths, countertops, and backsplash areas as indicated in the finish schedule

   c. Marble thresholds and dividers between tile and carpet or vinyl floor surfaces

   d. Mortar, thinset, mastic, leveling compound, and crack filler needed to complete the work

   e. Grout, rough-in, and pan in shower areas

2. Notify project superintendent of any walls out of plumb beyond 1/4 inch before installing tile. Install all cut pieces equally on two sides.
3. Do not cover up inferior or defective work.
4. Prepare surface and make minor repairs to substrate.
5. Install all work in accordance with specifications, procedures, and practices prescribed by the manufacturer.
6. Clean up all materials resulting from the work and remove from the site.
7. Clean all tile and remove stains, mastic, glue, thinset, and grout.
8. Protect all floor tile from damage.

Optional Scope Items:

9. Provide and install marble window sills.
10. Submit all tile samples and receive written approval before starting work.
11. Provide and hang cementitious backer board in all locations specifying ceramic tile (included in Finishes—Drywall).
12. Provide and hang moisture resistant sheetrock in all locations specifying ceramic tile (included in Finishes—Drywall).

Exclusions:

1. Molded, simulated marble countertops
2. Marble window sills

Finishes—Cementitious Stucco

Scope:

1. Provide all labor, material, hand tools, equipment (including safety equipment), supervision, supplies, and other incidentals necessary to install the stucco work and perform all such work in accordance with the contract drawings, specifications, addenda, and other documents that make up the contract documents but not limited to the following:

   a. Cementitious stucco, lath, and plaster

   b. Channel, furring, mesh, expansion joint, construction joint, stop molding, corner bead, vents, fasteners, and other accessories necessary to provide a complete installation

2. Protect all surfaces before application including aluminum windows and door trim to prevent etching.

3. Install work according to the specifications, procedures, and practices prescribed by the manufacturer(s).
4. Clean all excess stucco from windows, doors, trim, floors, patio slabs, and sidewalks.

Optional Scope Items for Multifamily and Light Commercial Projects:

5. Provide a sample board and receive written approval from builder before starting installation work.

Exclusions:

1. Synthetic stucco
2. Exterior Insulated Finish System (EIFS)

Finishes—Synthetic Stucco (Exterior Insulated Finish System) _____

Scope:

1. Provide all labor, material, hand tools, equipment (including safety equipment), supervision, supplies, and other incidentals necessary to install the synthetic stucco work and perform all such work in accordance with the contract drawings, specifications, addenda, and other documents that make up the contract documents but not limited to the following:
   a. Rigid board insulation, tape, mesh, joint materials base coat, and finish coat
   b. Flashing at roof to wall with kick-outs at the lowest end to prevent water from migrating behind wall finish
   c. Sealant for all expansion joints, window and door edges with backer rod, and a continuous seal to prevent water migration into the wall cavity
   d. Stop molding, corner bead, vents, fasteners, and other accessories necessary to provide a complete installation
2. Protect all surfaces before application including window and door trim to prevent etching. Cold weather protection to include tent and heat or termination of work until air temperatures rise to acceptable levels.
3. Install work according to the specifications, procedures, and practices prescribed by the manufacturer(s).
4. Allow base coat to cure before applying top coat.
5. Clean all excess materials from windows, doors, trim, floors, patio slabs, and sidewalks.

Optional Scope Items for Multifamily and Light Commercial Projects:

6. Provide a sample board and receive written approval from builder before starting installation work.
7. Provide an affidavit indicating that the installation has been made in accordance with the EIFS manufacturer's specifications.

Exclusions:

1. Cementititious stucco

Caution:

Because of the problems that have arisen with many homes with EIFS systems, NAHB is advising builders to carefully research the facts about these products before deciding to use them. For additional information about technical issues involving EIFS homes, builders can call the NAHB Research Center's HomeBase Hotline at (800) 898-2842.

At the time this book went to print a number of lawsuits were underway arising out of the use of EIFS. Some liability insurance carriers had informed builders that homes built with EIFS would not be covered under existing policies, and at least one bank had indicated that it would not provide mortgages on EIFS homes.

A July 11, 1997 article in NAHB's *Nation's Building News* reported "considerable debate over whether the traditional 'barrier' EIFS is compatible with other systems in a typical house and with the methods builders normally use to build homes." Builders should be cautious when using EIFS, particularly in relation to the risk of possible water migration around windows and in walls. A lot of these products are supposed to be installed by manufacturer-approved installers. However, while caulking and sealing are part of the manufacturers' specifications, the installers generally do not perform these tasks. Confusion over who is responsible for applying caulking and ensuring proper sealing has resulted in a lot of the water-related problems.

Product manufacturers have developed new drainable systems, some of which are now available as an alternative to the barrier EIFS. Additional EIFS products and systems are currently being evaluated by the code evaluation services.

Builders wishing to use EIFS must become familiar with the specifications and requirements of the particular product they are using. Components from different EIFS products should not be mixed. And builders must inspect the work thoroughly, both to ensure that the product has been installed according to manufacturer's specifications and to ensure that all potential areas of water infiltration have been flashed, caulked, and sealed. Builders who require their trade contractors to provide affidavits indicating that the installation has been made in accordance with the EIFS manufacturer's specifications should include a copy of the affidavit with documents used to register the project with the manufacturer for purposes of activating the EIFS manufacturer's warranty.

When writing a scope of work for EIFS installation, builders are urged to emphasize quality installation, well-drawn details, and adherence to the manufacturer's specifications. It is important to spell out who will be responsible for each step in the installation, especially who is responsible for the caulking, flashing, and sealing around all of the penetrations in the building envelope. It is also important to specify that substitution of components and materials from different manufacturers will require reinstallation with proper materials, the cost of which is to be borne by the trade contractor.

Finishes—Custom Millwork Wood Stairs

Scope:

1. Provide all labor, material, hand tools, equipment (including safety equipment), supervision, supplies, and other incidentals necessary to install all custom millwork stairs and perform all such work in ac-

cordance with the contract drawings, specifications, addenda, and other documents that make up the contract documents and not limited to treads, risers, nosing, wall stringers, moldings, handrails, balusters, easing, and newel posts.
2. Fabricate all work and install plumb, true, square, and level; properly attach to the building structures.
3. Measure wall and floor areas adjacent to stairs to verify dimensions prior to fabricating any materials.

Optional Scope Items:

4. Provide engineering for code compliance and shop drawing for written approval before starting work.

Exclusions:

1. Materials paid for by others

Finishes—Cabinetry

Scope:

1. Provide all labor, material, hand tools, equipment (including safety equipment), supervision, supplies, and other incidentals necessary to install the cabinets and perform all such work in accordance with the contract drawings, specifications, addenda, and other documents that make up the contract documents but not limited to the following:
   a. Custom and premanufactured cabinets
   b. Base and upper kitchen cabinets, vanities, utility, wet bar, and bookcases as indicated on the drawings
   c. Trim molding required at joints, corners, transitions, and cabinet surfaces that meet drywall to eliminate cracks in excess of 1/8 inch
   d. Laminated countertops and backsplashes
2. Caulk all cracks less than 1/8 inch to match the cabinet or wall.
3. Field measure to assure proper fit of all work after drywall is hung.
4. Clean up all materials resulting from the work and remove from site.
5. Install all work in accordance with specifications, procedures, and practices prescribed by the manufacturer(s).

Optional Scope Items:

6. Protect all countertops included in this scope until owner's acceptance.

Exclusions:

1. Molded, simulated marble countertops

Finishes—Interior Cleanup _____

Scope:

1. Provide all labor, material, hand tools, equipment (including safety equipment), supervision, supplies, and other incidentals necessary to install the final cleanup work and perform all such work in accordance with the contract drawings, specifications, addenda, and other documents that make up the contract documents but not limited to the following:
   a. Cabinets inside and out including kitchen, bathroom, wet bar, and utility room
   b. Countertops
   c. Appliances
   d. Range hoods
   e. Fixed glass, mirrors, windows inside and out, window frames, sliding mirror doors
   f. Interior doors, exterior doors, and sliding glass doors inside and out including tracks and frames, and door thresholds
   g. Plumbing fixtures including sinks, lavatories, tubs, showers, and water closets
2. Remove all paint spots, glue, caulking, and labels (except energy labels on appliances).
3. Remove cardboard and papers from within appliances. Do not remove energy labels. Leave all operating instructions in a drawer in the kitchen cabinets.
4. Prevent any damages resulting from tools, solvents, cleaners or chemicals used to complete this work.
5. Notify the project superintendent of any damaged materials (glass, plastic laminated cabinets, paint, mirrors).

Optional Scope Items:

6. Determine jointly with paint and drywall trade contractors and project superintendent need for excessive paint or texture coating over-spray.

Exclusions:

1. Site cleanup

Flooring—Carpet, Vinyl, and Wood _____

Scope:

1. Provide all labor, material, hand tools, equipment (including safety equipment), supervision, supplies, and other incidentals necessary to install the flooring work and perform all such work in accordance with the contract drawings, specifications, addenda, and other documents that make up the contract documents but not limited to the following:

    a. Carpet, pad, tack strips, metal edges, and accessories for a complete job (If a carpet is not selected, include a square yard allowance amount.)

    b. Resilient flooring sheet and tile materials including vinyl, vinyl composition tile (vct), vinyl base, and accessories for a complete installation (If a resilient flooring is not selected, include a square foot allowance amount.)

    c. Bonding coating, adhesives, glue, and mastic to secure the work

    d. Wood flooring, block, plank, and parquet flooring

    e. Related trim

    f. Wood sanding and finishing

2. Prepare floors, fill all score lines, cracks, and rough areas that are noticeable to the sight or feel when walking over the finished surface.

3. Install carpet with a tight fit, run with same direction; seam only when necessary.

4. Remove all glue, mastic, and stains, and protect finished surface.

5. Install all work in accordance with specifications, procedures, and practices prescribed by the manufacturer.

6. Include an underlayment if recommended by the manufacturer.

7. Caulk vinyl flooring around plumbing fixtures.

8. Cut in all vinyl to fit edges around walls, fixtures, and cabinets to prevent gaps.

9. Seal all vinyl seams.

10. Clean up all materials resulting from the work and remove from the site.

Optional Scope Items:

11. Submit all carpet samples and receive written approval before starting work.

Exclusions:

None

Appliances

Scope:

1. Provide all labor, material, hand tools, equipment (including safety equipment), supervision, supplies, and other incidentals necessary to install the appliances and perform all such work in accordance with the contract drawings, specifications, addenda, and other documents that make up the contract documents but not limited to the following:

    a. Ranges, refrigerators, ice makers, washers, dryers, and required pigtails and cords needed to connect appliances to utilities

    b. Venting pipe, dampers, caps, and related hardware to connect the appliances to the outside air as required by the drawings

    c. Ice maker line to water supply

2. Deliver all appliances inside units. If appliances are not selected, include a total allowance amount.
3. Set up and level appliances.
4. Remove all boxes and trash from site.
5. Install all appliances in accordance with codes. If tailgate delivery was specified and no installation is included, be sure to have vendor provide a time and date for delivery.
6. Hook up water line to ice maker.
7. Clean up all materials resulting from the work and remove from the site.

Exclusions:

1. Rough-in appliance venting
2. Installation of dishwasher, cooktop, built-in oven, disposal, and other hard-wired appliances

Prefabricated Metal Fireplace _____

Scope:

1. Provide all labor, material, hand tools, equipment (including safety equipment), supervision, supplies, and other incidentals necessary to install prefabricated metal fireplaces and perform all such work in accordance with the contract drawings, specifications, addenda, and other documents that make up the contract documents but not limited to the following:
   a. Wood burning and gas fireplaces
   b. Flues, firestops, bracing, flue offsets, chimney caps, glass doors, combustion venting, and incidentals needed for complete working fireplaces
2. Install all work in accordance with specifications, procedures, and practices prescribed by the manufacturer.
3. Coordinate installation of all required firestops and draftstops with drywall contractor and building officials.
4. Assure design height of chimney is in accordance with the manufacturer's specifications.
5. Clean up all materials resulting from the work and remove from the site.

Exclusions:

1. Masonry components for the fireplace
2. Metal firebox

Heating, Ventilating, and Air Conditioning _____

Scope:

1. Provide all labor, material, hand tools, equipment (including safety equipment), supervision, supplies, and other incidentals necessary

to install the heating, ventilating, and air conditioning (HVAC) work and perform all such work in accordance with the contract drawings, specifications, addenda, and other documents that make up the contract documents. The intent is to provide and install a complete heating and air conditioning system including but not limited to the provision and installation of the following:

a. Mechanical equipment, ductwork, hangers, sleeve frames, flashing, supports, insulation, and roof curbs

b. Control systems, control wiring, thermostats, and all incidentals for a complete job

c. Condensation drain piping and traps

d. Low-voltage wiring needed to operate the system

e. Gas venting, vent pipe, caps, transitions for heating system, and gas hot water tank

f. Pressure testing of gas lines

2. Coordinate the location and installation of ducts, freon, and condensation lines with other trades to insure a free and unencumbered area for the runs.

3. Coordinate with the drywall trade contractor to insure that all return air plenums are properly and completely sealed and free of air loss.

4. Start up, test, and balance the entire system.

5. Clean up all materials resulting from the work and remove from the site.

6. Install specified range and cooktop exhaust and bath venting.

Optional Scope Items:

7. Install required housekeeping equipment pads for mechanical equipment.

8. Cut, core, and patch as required for this work.

Exclusions:

1. Mechanical equipment pads

2. High-voltage wiring

Fire Protection

Scope:

1. Provide all labor, material, hand tools, equipment (including safety equipment), supervision, supplies, and other incidentals necessary to install the fire protection work and perform all such work in accordance with the contract drawings, specifications, addenda, and other documents that make up the contract documents. The intent is to provide and install a complete fire sprinkler system and includes but is not limited to the complete installation of the following:

a. Sprinkler piping (main and laterals), disconnects, hookups, headers, drops, heads, valves, valve pits, hydrants, stand pipe, automatic sprinklers, and related special fixtures necessary to complete the installation in the areas required by codes

    b. Flushing, purging, testing, and adjusting as required to place the system into final service

2. Coordinate the installation of this work with interfacing trades.
3. Coordinate with the local fire inspector any particular sequence or requirements in the installation of the work.
4. Provide all required governmental approvals.
5. Pressure test prior to the work being covered up by other trades.

Optional Scope Items:

6. Cut, drill, core, and patch as required for the installation of this work.

Exclusions:

None

Plumbing

Scope:

1. Provide all labor, material, hand tools, equipment (including safety equipment), supervision, supplies, and other incidentals necessary to install the plumbing work and perform all such work in accordance with the contract drawings, specifications, addenda, and other documents that make up the contract documents but not limited to the following:

    a. Water lines, sanitary sewer lines stubbed out 5 feet from building, and connections to mains

    b. Piping, valves, backflow preventors, vents, floor drains, roof drains, sleeves, hangers, supports, insulation, and accessories needed to complete the work

    c. Fixtures including lavatories, water closets, tubs, sinks, and trim (If fixtures are not selected, include an allowance amount.)

    d. Equipment specified including water heaters, fixtures, stops, valves, and trim

2. Pressure test lines and connections before they are covered up.
3. Install dishwasher and disposal.
4. Obtain permits for the work.
5. Clean up all materials resulting from the work and remove from the site.

Optional Scope Items:

6. Connect sanitary sewer to main.
7. Connect water lines to main.
8. Install housekeeping equipment pads for mechanical equipment.
9. Cut, drill, core, and patch as required for this work.
10. Trench, backfill, and compact as required to perform the work.

11. Install temporary water lines and bibs necessary for construction.
12. Install gas lines and meter.

Exclusions:

1. Site water lines
2. Water meter
3. Gas meter
4. Tap fees
5. Shower pan
6. Sewer lateral from house to utility line or septic tank
7. Septic system
8. Condensation piping for condensation water

Electrical

Scope:

1. Provide all labor, material, hand tools, equipment (including safety equipment), supervision, supplies, and other incidentals necessary to install the electrical work and perform all such work in accordance with the contract drawings, specifications, addenda, and other documents that make up the contract documents. (If the drawings are not complete, consider including a phrase to comply with codes instead of a reference to the drawings.) The intent is to provide and install a complete electrical system and includes but is not limited to the complete provision and installation of the following:
   a. Individual panels, subpanels, breakers, switches, conductors, and conduit for each metered system including circuits for heating, air conditioning, appliances, water heater, lighting, GFIC, receptacles, switches, and accessories
   b. Primary and secondary conductors, conduits, and grounding
   c. Light fixtures, fans, range hoods, smoke detectors, bulbs, and lamps (If fixtures are not selected, include an allowance amount.)
   d. Motor power wiring between the energy source and the motor
   e. Emergency systems, security systems, communications systems, fire alarm systems, and detection systems
   f. Temporary power poles, installed in such a manner that trades will not have to pull more than 150 feet of extension lines to perform their work
   g. Television and telephone systems necessary to have working television and telephone outlets
2. Color code and mark as required.
3. Install all hard-wired appliances including cooktop, microwave oven, built-in oven, and range hood.
4. Coordinate with the power company and other trades to interface work.

5. Coordinate with applicable cable and telephone utility companies.
6. Test and hot check entire system.
7. Clean up all materials resulting from the work and remove from the site.

Optional Scope Items:

8. Provide and install building panels, subpanels, breakers, switches, conductors, and conduit.
9. Provide and install exit and emergency lights.
10. Provide and install temporary wiring for the project superintendent's office and building.
11. Provide and install plywood backboards for a temporary telephone.
12. Provide and install cable tray and bus duct systems shown on drawings.
13. Provide and install uninterrupted power supply, switchgear, and switch equipment.
14. Provide and install all control wiring required for automatic starting and stopping of motors specifically shown on the electrical drawings.
15. Cut, core, and patch as required.
16. Trench, backfill, and compact as required for this scope.

Exclusions:

1. Site electric
2. Appliances

Chapter 4

Steps Leading to the Trade Contract

Successful trade contracting does not happen by accident. It requires careful, detailed planning followed by skillful orchestration of multiple trade contractors. Start your contract planning by making a detailed analysis of the project and identifying all construction tasks. Then decide which tasks your work force will execute and which will be done by trade contractors. Prepare trade contracts for all work that is to be contracted.

Before you send out requests for quotes, make a list of reliable, quality trade contractors. We recommend that you request a quote from at least three contractors to ensure competition. Price should not be the only criterion for selection of trade contractors. As you develop your list of prospective contractors, consider their past responses to warranty calls, reputation for quality work, and commitment to providing OSHA-compliant, safe jobs. Exercise the same care when selecting trade contractors that you would use when hiring new employees.

Treat trade contractors with respect and adequately compensate them for the work they perform. Strive to develop long-term relationships with the best trade contractors to ensure well-executed projects and satisfied homeowners. Remember that much of your reputation for quality projects is based on work actually performed by the trades.

At the end of the planning process, send requests for quotes to the prequalified, prospective trade contractors on your list.

Develop a Contracting Strategy

The first step toward developing a contracting strategy is to analyze the project completely and identify all the work packages required for its successful completion. A work package consists of one or more activities that must be completed to finish the project. A sample breakdown of the work packages for the Smith residence, discussed in Chapter 2, is shown in Figure 4-1. The size of each work package and the number of packages depends on the scope and complexity of the project. It is crucial that all elements of work be identified when preparing the trade contracts. The level of detail used to identify work packages is the builder's choice.

Once the work packages are identified, builders must make three decisions:

Work Packages for Smith Residence

1.	Sitework	12.	Cabinets
2.	Footings	13.	Countertops
3.	Foundation Walls	14.	Paint and Stain
4.	Flatwork	15.	Floor Covering
5.	Rough Carpentry	16.	Specialties
6.	Finish Carpentry	17.	Gas Furnace
7.	Siding System	18.	Appliances
8.	Roofing System	19.	Mechanical
9.	Insulation	20.	Miscellaneous Venting
10.	Doors and Windows	21.	Plumbing
11.	Gypsum Wall Board	22.	Electrical

Figure 4-1. Sample Work Breakdown

- Which work packages their crews will perform
- Who are the available, qualified trade contractors
- How much risk they are willing to incur on the project

Based on these decisions, builders select which work packages to contract and which to perform using their own work crews. Contracting most of the work to qualified trade contractors allows builders to minimize their risk during project execution as discussed in Chapter 1. If builders cannot identify reliable, qualified trade contractors, however, they may choose to perform the work themselves to ensure a quality product. This may involve hiring additional workers who will require supervision.

The process of identifying and selecting work packages that will be contracted and determining the individual work packages or combinations of work packages for each trade contract is called developing the contracting strategy. In our example of the Smith residence, the builder, Cascade Homes, developed the contracting strategy shown in Figure 4-2. In this example, the builder employs its own carpentry crews, but uses trade contractors for all other work.

Select the Contract Scope of Work

After identifying the work packages, decide whether each work package requires a separate contract or whether some of the packages can be combined into a single contract. For example, the concrete foundation and concrete driveway might be combined in a single contract.

Contracting Strategy for Smith Residence

Work Package	Method of Performance
1. Sitework	Trade Contractor
2. Footings	Trade Contractor
3. Foundation Walls	Trade Contractor
4. Flatwork	Trade Contractor
5. Rough Carpentry	Cascade Homes
6. Finish Carpentry	Cascade Homes
7. Siding System	Trade Contractor
8. Roofing System	Trade Contractor
9. Insulation	Trade Contractor
10. Doors and Windows	Trade Contractor
11. Gypsum Wall Board	Trade Contractor
12. Cabinets	Trade Contractor
13. Countertops	Trade Contractor
14. Paint and Stain	Trade Contractor
15. Floor Covering	Trade Contractor
16. Specialties	Trade Contractor
17. Gas Fireplace	Trade Contractor
18. Appliances	Trade Contractor
19. Mechanical	Trade Contractor
20. Miscellaneous Venting	Trade Contractor
21. Plumbing	Trade Contractor
22. Electrical	Trade Contractor

Figure 4-2. Sample Contracting Strategy

Next, determine which jobsite management responsibilities will be assigned to each trade contractor. For example, the mason might be required to leave the scaffolding for the stucco contractor.

Finally, prepare a detailed scope of work for each trade contract using the model scopes of work provided in Chapter 3, as needed, to suit the individual project. In our example of the Smith residence, the builder decided to combine miscellaneous venting and mechanical work in a single trade contract. A separate trade contract was developed for each of the other work packages that were contracted. The complete scope of work for the electrical contract is shown in Figure 2-1.

Define Contract Terms and Conditions

After developing the scopes of work, prepare the terms and conditions for each trade contract as discussed in Chapter 2. You can use a standardized set of terms and conditions for all trade contracts on a specific project. Include any unique requirements in the individual contracts' scope of work. Some builders choose to use a standardized set of terms and conditions for all trade contracts awarded by their company. The sample electrical contract shown in Chapter 2 combines terms and conditions specific to the trade with general responsibilities and a standardized set of terms and conditions that are included in the basic contract by reference. Select a document system that best serves your needs.

Identify Potential Trade Contractors

Although many builders use the same set of trade contractors for their projects, it is good business practice to seek to enlarge your network of reliable trade contractors. The objective is to develop a list of reliable specialty contractors who are interested in working on your projects and who work well with the other trades. Over time, they will become familiar with your preferences. Recruit new trade contractors before you actually need them. Most trade contractors work for multiple builders and might have difficulty meeting your schedule requirements without adequate notice.

Networking through the local builder's association, with other builders, with local suppliers familiar with trade contractors, and with current trade contractors are all good ways to identify reliable trade contractors. Another good source are local building inspectors who know which trade contractors produce quality work. Simply driving around sites being built by quality builders also can provide a wealth of information regarding the capabilities of local trade contractors.

Prequalify Potential Trade Contractors

To help ensure quality work, consider prequalifying all trade contractors before you ask them to submit a quote. Evaluate each of the trade contractors using a standard set of criteria:

- Experience and technical skills related to work required
- Technical and supervisory competency of management and field supervisors or crew leaders
- Stability and financial strength of company (e.g., assets, liabilities, debt load, and time in business)
- Adequacy of equipment and labor (e.g., well trained crews, availability, particularly if the contractor works for several builders, and flexibility given potential changes in schedule)

- Safety performance including past workers' compensation claims, OSHA violations or fines (if any), and existing safety and health programs
- Reputation for working with builders and other trades, proven history of responding to warranty calls, and past contract performance

One way to gather much of this information is to require prospective trade contractors to respond to a detailed questionnaire. A sample questionnaire is shown in Figure 4-3 and duplicated on the companion diskette. Do not rely solely on the data provided on the questionnaire. Contact other builders who have contracted with the prospective trade contractor and check the contractor's credit with local credit associations and suppliers. Contractors who do not pay suppliers in a timely manner might be unable to procure materials when needed and delay the project.

Some small builders believe contractor prequalification is not necessary because of the size of their business or because the contractors they typically hire consist of one person with a pickup truck. Prequalification, however, is always a good practice when seeking trade contractors.

Request Quotes from Trade Contractors

Although many builders get unit price quotes from the same set of trade contractors for each project, prudent builders constantly check the market to make sure they are receiving competitive quotations for each trade contract. It is good business practice to obtain at least three quotes for each trade contract to ensure competitive pricing.

Some builders solicit quotations by telephone, but we recommend using written requests for quotes to ensure that each prospective contractor understands the project requirements. For larger projects, the solicitation package needs to contain a detailed scope of work, the terms and conditions of the contract, and the drawings and specifications. If contract drawings and specifications are not provided with the request for quotes, prospective contractors must be told where they can be obtained. On small projects, a one-page request containing a detailed scope of work attached to a drawing and specifications might be all that is needed.

It is also a good idea to provide trade contractors with a format for submitting their quote and a date when the quote is due. A request for the Smith residence electrical contract is shown in Figure 4-4. The terms and conditions shown in Figure 2-1 are to be attached to the request in Figure 4-4 when it is given to the electrical contractor.

Preparing a set of requests for quotes for 20 trades may seem like a great amount of work, but with a computer and word processing software they can be generated from a template set in an hour or less.

Cascade Homes

1650 Happy Valley Road
Olympia, Washington 98507

TRADE CONTRACTOR QUESTIONNAIRE

1. General Company Information

Name:_____

Address:_____

City:_____ State:_____ Zip:_____

Telephone:_____ FAX:_____

(Check appropriate box.)

❑ Sole owner. Name:_____

❑ Partnership. List names of partners:

_____

_____

_____

_____

❑ Corporation.

President:_____

Vice President:_____

Number of years in business under present name:_____

Continued

Figure 4-3. Sample Trade Contractor Questionnaire

1. **General Company Information** (continued)

   Trade(s) normally performed by company: _____

   _____

   Number of trade and office personnel currently employed:

   Trade employees: _____   Office employees: _____

2. **Financial Information**

   What is the maximum dollar value of work the company is capable of handling at one time?

   $ _____

   Attach last 2 years of audited financial statements at end of questionnaire.

3. **Insurance Information**

   What is the company's workers' compensation experience modification rate for the 3 most recent years?   19____ _____   19____ _____   19____ _____

   How much insurance coverage does the company currently carry?

	Yes	No	Amount
General Liability	___	___	_____
Automobile Liability	___	___	_____
Workers' Compensation	___	___	_____

4. **Safety Information**

   Does the company have a written safety program?   _____ Yes   _____ No

   Use your OSHA Form 200 to complete the following table.

	19____	19____	19____
Total Number of Workers' Compensation Claims			
Number of Lost Time Workers' Comp. Claims			
Number of Accident Liability Claims			
Number of Fatalities			

   Continued

Figure 4-3. Sample Trade Contractor Questionnaire (*continued*)

5. Project Information

List current, ongoing projects with approximate dollar value and estimated completion date.

Project	Amount	Completion Date

Has the company failed to complete any work assigned to it during the past 5 years?

_____ Yes   _____ No

If Yes, explain:

_____

_____

_____

Attach a list of projects completed in the last 3 years and a list of builder references whom we may contact.

6. Equipment Information

Attach a list of owned construction equipment with capacity, age, type, and attachments.

This questionnaire was completed by:

Name:_____ Title:_____

Signature:_____ Date:_____

Figure 4-3. Sample Trade Contractor Questionnaire *(continued)*

Cascade Homes

1650 Happy Valley Road
Olympia, Washington 98507

REQUEST FOR QUOTATION

To: Mountain Electrical Contractors   RFQ # S-16
     1632 West Pacific Avenue   Date: May 3, 19__
     Fife, Washington 98424

Project: Smith Residence

Project Address: 7654 Mountain View Road
                                  Tacoma, Washington 98411

Quote Required by: May 27, 19__

Contract Documents: General contract for construction of Smith Residence prepared by Northwest Architects, Inc.

Scope of Work: Provide all materials, labor, tools, equipment, supervision, supplies, and other items necessary or required to safely execute the work and perform all electrical work in accordance with the contract drawings, specifications, and any addenda contained in the general contract documents for the construction of the Smith residence. The intent is to provide a complete electrical system that meets local code requirements including, but not limited to, provision and installation of the following:

1. Primary and secondary conductors, conduits, and grounding
2. All panels, subpanels, breakers, and switches for each system including
   circuits for: heating, air conditioning, appliances, water heater, lighting, GFCI,
   receptacles, switches, and accessories
3. Light fixtures, fans, range hood, smoke detectors, bulbs, and lamps
4. Color coding and marking, as required
5. Cutting, coring, and patching, as required
6. Test and hot-check of entire electrical system

Contractor's quote will be based on the complete set of contract documents that are available at Cascade Homes' business office.

Terms and conditions of the electrical contract are attached.

Continued

Figure 4-4. Sample Request for Quotation

To: Cascade Homes
1650 Happy Valley Road
Olympia, Washington 98507

We propose to perform the entire scope of work for: _____

We agree to perform the described work for the quoted price if notified of acceptance by Cascade Homes within 60 days of date submitted.

Subcontractor: MOUNTAIN ELECTRICAL CONTRACTORS

By: ➢_____
Authorized Signature/Title

Date: _____/_____/_____

Figure 4-4. Sample Request for Quotation *(continued)*

Builders who commit to using written documents have better control over their business.

Estimate Cost for Each Contract Scope of Work

Builders should develop their own cost estimate for each trade's scope of work to ensure the quotes received are reasonable. When developing the cost estimate, use the same methodology the trade contractor will use to prepare the quote. This gives you a better understanding of the trade contractor's concerns when he or she prepares the quote and makes comparison easier.

Your cost estimate is an important cost control tool. Use it to ensure that contracted scopes of work can be completed within the overall project budget and that quotes are reasonable. For more information on residential construction estimating, contact the Home Builder Bookstore at the National Association of Home Builders. The bookstore carries several publications presenting various approaches to estimating.

Gather and Evaluate Trade Contractor Quotes

As discussed earlier in this chapter, we recommend that builders obtain several quotes for each trade contract. An efficient way to compare quotes both to each other and to your own cost estimate is to prepare a quote evaluation sheet such as the example shown in Figure 4-5. A blank form is included in the Appendix and duplicated on the companion diskette.

Review all quotes for conformance with the request for quote to ensure contractors did not leave out any items or add some that are included in another contract's scope of work. Compare the quote to your cost estimate to ensure the prices are reasonable. Unrealistically low or excessively high quotes generally indicate that the trade contractor:

- Was uncertain about the exact scope of work
- Found ambiguities regarding some of the conditions of work

Cascade Homes

1650 Happy Valley Road
Olympia, Washington 98507

QUOTATION EVALUATION SHEET

Project: _LOT 30, HIDDEN COVE_

Contract Scope: _PLUMBING_

	COST
Estimated Cost: *by BOB JONES*	$8,000.00
CONTRACTOR	
JAY'S PLUMBING	$8,200.00
NORTHWEST MECHANICAL	$9,500.00
A-1 PLUMBING	$8,350.00

Figure 4-5. Sample Quotation Evaluation Sheet

- Erred when preparing the quote
- Changed specifications
- Omitted items

Ambiguities may be resolved by telephone and then followed up in writing. If extensive clarification is needed, however, you might need to solicit new quotes. In some cases, you may negotiate the price with the trade contractor. This might involve modifying some of the elements in the scope of work to reach a mutually agreeable price. Always follow up verbal agreements in writing.

Select Trade Contractors

After evaluating the quotes, select the trade contractors that represent the best value based on three evaluation criteria:

- Business competency
- Technical competency
- Proposed price

Remember that the lowest price does not necessarily represent the best value.

Since performance bonds generally are not used in residential construction, it is crucial that the builder ensure that each selected trade contractor is a viable business enterprise with adequate financial resources to complete the job. Likewise, the contractor's technical competence and ability to deliver a quality product in the desired time frame is a crucial selection criteria. The quoted price still is a major consideration, but you also should take into account the trade contractor's crew availability and expertise, ability to meet the projected schedule, acceptance as a team member by other trades on the jobsite, and his or her reputation for quality of work.

Select trade contractors that represent the best overall value. The lowest price may end up being the most expensive option. The decision to pay more for quality workmanship often results in a quality reputation that attracts additional work since a builder's best marketing assets generally are quality homes and satisfied homeowners. Carefully select quality craftsmen, even at premium cost, if you want to earn a reputation for quality projects. When possible, actually examine samples of the contractors' work before you select them for your projects.

Award the Trade Contracts

Once you select the group of trade contractors for your project, award the contracts. A contract award represents acceptance of the contractor's quote as submitted or as revised through negotiations. To create a contract, use the scope of work and contract terms and conditions developed for the request for quote and add the contract price. An example of a completed trade contract is shown in Figure 2-1 and included on the diskette.

Notify the trade contractors that you do not decide to use. Tell them about the process you used to make your decisions and why they were not selected. This educates them as to your needs and hopefully will make them more competitive. Ultimately, this might provide you with more quality contractors from which you can build a stronger team.

Chapter 5

Managing Trade Contractors

Once trade contracts are awarded and construction begins, the builder's efforts shift to coordinating the contractors' work to ensure the project is completed on time, within budget, and in accordance with contract requirements. Since most of the work is performed by the trades, efficient management of trade contractors is crucial to a builder's ability to control project costs and complete the project on time. In this chapter, we discuss some of the major issues involved in developing management systems to ensure a profitable project.

The builder must communicate frequently with trade contractors to apprise them of changes in the construction schedule and anticipated changes in their scope of work. This usually is done by telephone or on the jobsite. The builder also must schedule frequent inspections of their work to ensure conformity with contract requirements.

Build a Team

The most fundamental management challenge builders face when constructing a project is molding trade contractors into a cohesive team. This requires an understanding of the trade contractors' concerns and the ability to organize the project. A good starting point is weekly meetings with trade contractors to discuss their concerns and solicit advice regarding project execution.

Frequent and open communication coupled with mutual goals will help you develop relationships with the trades that are built on trust and a spirit of partnership. Public recognition of individual craftspeople also enhances the notion that the trades are part of your team. In some cases, trade contractors need to be trained regarding your project management practices, materials security, safety program, and quality standards. Training may take the form of periodic seminars or non-project training sessions. The objective is to foster long-term relationships with the trades that are profitable for everyone.

Control the Schedule

The success of any building project is dependent on a viable schedule. This is especially important when you use numerous trade contractors to complete the project. It is the responsibility of the builder to coordinate the tasks contracted to the trades so that start dates are realistic. An effective schedule promotes the team spirit needed to ensure a quality project.

Develop the Schedule

The project schedule identifies the relationship and sequence of all activities and the amount of time allowed for the completion of each. The schedule must be realistic and take into consideration working conditions, special milestone dates, and weather. Trade contractor input is essential to the development of an effective schedule. Use the schedule to coordinate the construction work and identify activities that are falling behind, coordinate the project, and complete each phase of the work on time. A sample master construction schedule for a residential project is illustrated in Figure 5-1. It shows the scheduled start and completion times for all trade contractors' activities.

			July 1996
Description	**Resp**	**Duration**	Cal Days: 8 9 10 11 12 15 16 17 18 19 22 23 24 25 26 29 30 31 / Workdays: 1 2 3 4 5 6 7 8 9 10 11 12 13 14 15 16 17 18
Slab Prep Work	Concrete Crew	4 days	▓▓▓▓ (Workdays 1–4)
Inspect Slab	Inspector	1 day	▓ (Workday 5)
Place & Finish Slab	Concrete Crew	1 day	▓ (Workday 6)
Erect Wood Framing	Framing Crew	5 days	▓▓▓▓▓ (Workdays 7–11)
Comp Rough Framing	Framing Crew	2 days	▓▓ (Workdays 12–13)
Install Roof Shingles	Roofing Crew	2 days	▓▓ (Workdays 16–17)
Inst Door & Windows	Framing Sub	1 day	▓ (Workday 12)
Inst Siding & Cornice	Siding Sub	2 days	▓▓ (Workdays 13–14)
R/I & Trim Electrical	Electrical Sub	2 days	▓▓ (Workdays 15–16)
Paint Exterior	Painting Sub	3 days	▓▓▓ (Workdays 15–17)
Final Inspection	Inspector	1 day	▓ (Workday 18)

Figure 5-1. Sample Project Schedule

From Thomas A. Love, *Scheduling Residential Construction for Builders and Remodelers* (Home Builder Press, 1995), 13.

Notify the Trades

Trade contractors need adequate time to arrange their scheduled jobs, obtain equipment and materials, and schedule their crews. This means the builder must provide adequate notice to each trade contractor regarding the scheduled start time for his or her phase of work. In residential construction, this notice usually is communicated by telephone. Time notice is fundamental to building good relationships with trade contractors. The notice also should indicate the scheduled completion date to eliminate interference with the next trade's work.

If different trade contractors will work concurrently on the site, the builder should ensure that the trades are compatible and that each receives notice. Whenever the builder schedules simultaneous or dovetailing work, careful orchestration is crucial to ensure timely completion of the project. All contractors should be aware of their safety responsibilities including those toward other contractors who may also be onsite. Required safety equipment and devices should be maintained and made available as needed to comply with the law and ensure adequate protection of workers onsite.

Failure to update the contractors regarding project status or allowing a contractor to arrive on a jobsite that is not ready causes a hardship for the trade contractor and can result in lost time and profit for the builder. Proper notice is paramount to the smooth transition between the trades and their ongoing relationship with the builder. Trade contractors who cannot make a profit find other builders. Without quality trades, builders soon become noncompetitive.

Monitor Quality Control

Quality control must be the builder's priority and enforced from the start of construction. Indeed it is a good idea to monitor quality control before construction begins by ensuring that appropriate materials are used and contracts awarded to the best trade contractors. Establish quality standards and make sure the entire construction team actively enforces them. Do not allow trade contractors to build over work that has been done improperly by a previous trade.

Schedule inspections as milestones to monitor progress and check standards. We recommend the use of quality-control checklists to ensure systematic inspections and record their results. Such checklists are described in *How to Hire and Supervise Subcontractors,* by Bobby R. Whitten, published by Home Builder Press.

Builders must ensure that each trade contractor understands the level of workmanship required in the contracted scope of work. You might want to see samples of work before you sign a trade contract. On high-end custom homes, you might want to have selected contractors construct mock-ups, which are stand-alone samples of completed work. Examples might be a small brick wall to demonstrate the masonry contractor's workmanship or samples of interior finishes. This allows you to evaluate the work and establishes a standard for the remaining work. It also allows homeowners to see what they are getting ahead of time and avoids misconceptions regarding types of finish.

It is good management practice to walk the project when each trade finishes its portion of the work to identify any needed rework. This is better than waiting until the end of construction and trying to get several trades to return and do rework. It also prevents poor work from being covered. Early corrective action provides the succeeding trades a clean start and improves expectations for a quality project. Another technique frequently used is to have follow-on-trades inspect the completed work of preceding trades. Discuss trade handoff criteria during weekly site meetings to ensure the trades are working together.

Record the results of formal inspections by the owner or architect and building departments and informal inspections with trade contractors in the daily job diary. Make the results of formal inspections a matter of record. Both the builder and trade contractor should keep copies of the results. Inspection results should indicate the date of the inspection, names of people present, portions of work inspected, and any deficiencies noted. A sample inspection report form is shown in Figure 5-2. A blank copy of the form is included in the Appendix and on the diskette.

Control the Contract

Even though the work on a project may be accomplished by a number of trade contractors, the builder retains responsibility for the successful completion of the project. It is the builder who must keep the work progressing and the homeowner and trade contractors happy. To do this effectively, builders must abide by both the general contract with the homeowner and the individual contracts with the trades.

Just as you develop schedules and job diaries, you should maintain documents that enable you to control the trade contracts. These controls are an important part of the financial management of the project and may prove valuable in legal disputes should the need arise.

Make Contractor Payments

Trade contractors usually submit invoices as their work progresses, at the end of each phase of their scope of work, or according to the payment schedule provided in the trade contract. Once the builder approves an invoice, payment is made to the contractor. Payments should be made within the timeline established in the trade contract, such as 15 days after receipt of the invoice. Pay the trades in a timely manner to avoid causing them cash flow problems and to ensure their financial health.

Keep copies of trade contractor's invoices and record the amounts actually paid. This documentation is essential for the financial management of the project. Information regarding contractor payments can be tracked online (e.g., input into a job-cost control system or scheduling system that tracks job costs) or simply entered in a ledger.

Lenders often require that a partial lien release be signed by the builder before a progress payment is made. Builders also should require a partial lien release from trade contractors to ensure they have paid all material bills. This makes it harder for vendors to lien the property. A sample partial lien release form is shown in Figure 5-3. Before making final payment to a trade contractor, obtain a final lien release to ensure liens will not be placed on the project. A sample final lien release form is shown in Figure 5-4. Blank copies of a partial and final lien release form are included in the Appendix and on the diskette.

Cascade Homes

1650 Happy Valley Road
Olympia, Washington 98507

PROJECT INSPECTION REPORT

Project: _LOT 30, HIDDEN COVE_   Date: _Nov. 10, 1997_

Inspection Participants:

BILL WALLACE, SUPERINTENDENT
BOB JONES, NORTHWEST FOUNDATIONS
KEN ADAMS, JAY'S PLUMBING

Scope of Inspection:

FOUNDATION WALLS AND FOOTINGS
UNDER SLAB PLUMBING

Items Requiring Correction:

1) ADD ½" x 8" ANCHOR BOLT BY ENTRY.

2) PATCH HOLE IN WALL BY NE CORNER.

3) CLEAN UP FORM MATERIAL.

Figure 5-2. Sample Inspection Report Form

PARTIAL LIEN RELEASE
UPON RECEIPT OF PROGRESS PAYMENT

Upon receipt by the undersigned of a check from _CASCADE HOMES_

in the sum of $_2,000.00_ payable to _JAY'S PLUMBING_
and when the check has been properly endorsed and paid by the bank upon which it is drawn, this document shall become effective to release any mechanic's lien rights the undersigned has on the project of:

LOT 30 HIDDEN COVE located at _RENTON, WASHINGTON_
to the following extent. This release covers a progress payment for labor, services, material, and equipment furnished to:

CASCADE HOMES through _NOV. 30, 1997_ only and
does not cover labor, services, material, or equipment not compensated by the progress payment.

Dated: _DEC. 10, 1997_   _JAY'S PLUMBING_

                               By: _Ken Adams_

Figure 5-3. Sample Partial Lien Release Form

Require Written Change Orders

Situations may arise where the scope of work for a trade contract needs to be modified. All such modifications should be documented as a contract change order to identify clearly the changes in scope and the impact on the trade contract price. Sometimes changes affect multiple trades. In these cases, all affected trade contracts must be changed appropriately. Notify all impacted trades of the change and then negotiate a contract adjustment to compensate the contractors for the impact.

Make all change orders in writing and have them authorized by the appropriate individual on the builder's project staff. Warn trade contractors that payment will be made only for changes properly authorized by the builder. Do not implement homeowner-requested changes until approved by the builder. Issue no-cost changes that modify the plans and specifications in writing to provide a record of changes to contract

FINAL LIEN RELEASE
UPON RECEIPT OF FINAL PAYMENT

Upon receipt by the undersigned of a check from _CASCADE HOMES_

in the sum of $ _3,000 ⁰⁰_ payable to _JAY'S PLUMBING_
and when the check has been properly endorsed and paid by the bank upon which it is drawn,
this document shall become effective to release any mechanic's lien rights the undersigned has
on the project of:

LOT 30, HIDDEN COVE located at _RENTON, WASHINGTON_.
The undersigned has been paid in full for all labor, services, material, and equipment furnished

to: _CASCADE HOMES_ .

Dated: _MARCH 1, 1998_   _JAY'S PLUMBING_

By: _Ken Adams_

Figure 5-4. Sample Final Lien Release Form

documents. Examples of such no-cost changes might be a new paint color or moving a window before the wall is framed.

We recommend using change order registers to document the changes to trade contracts. Annotate changes to plans and specifications on a master set of documents and cross-reference them to the change order number and date of change. A sample change order is shown in Figure 5-5 and a sample change order register is illustrated in Figure 5-6. Blank copies of both are included in the Appendix and on the diskette.

Terminate Contracts with Care

In the event the trade contractor performs poorly, the builder must issue and make a matter of record the proper written notice specified in the contract terms and conditions. Follow a first notice with a second, if necessary, to provide an opportunity to salvage the relationship.

Check with your attorney prior to terminating a contract. The need to terminate a contract can be minimized by carefully selecting quality

Cascade Homes
1650 Happy Valley Road
Olympia, Washington 98507

CONTRACT CHANGE ORDER

Change Order Number: _____ *1* _____

Project: _LOT 30, HIDDEN COVE_ _____

Contract: _____

We agree to make the following change(s) in the contract scope of work:

CHANGE TO BONE COLOR FIXTURES _____

_____

_____

Change(s) will affect the following plans and/or specifications:

_____

Subject to the following adjustment to the contract value:

Cost of this Change: _____ *$800.00* _____

Previous Contract Amount: _____ *$8000.00* _____

Revised Contract Amount: _____ *$8,800.00* _____

Revised Completion Date: _FEBRUARY 15, 1998_

Builder: CASCADE HOMES

By: ➤ _Bill Wallace_ _____ Date: _Nov. 1, 1997_
              Authorized Signature/Title

Subcontractor: _JAY'S PLUMBING_

By: ➤ _Ken Adams_ _____ Date: _Nov. 1, 1997_
              Authorized Signature/Title

Figure 5-5. Sample Contract Change Order

Cascade Homes

1650 Happy Valley Road
Olympia, Washington 98507

CHANGE ORDER REGISTER

Project: *LOT 30, HIDDEN COVE*

Trade Contractor: *JAY'S PLUMBING*

Date Awarded: *SEPT. 15, 1997*

				Contract Value
		Original Contract Value		*$8000.00*
Change Order Number	Date of Change Order	Scope of Change Order	Change Order Value	
1	*Nov 1, '97*	*ADD COLOR FIXTURES*	*$ 800.00*	*$8,800.00*
2	*Dec. 6, '97*	*ADD WHIRLPOOL TUB*	*$1,000.00*	*$9,800.00*

Fig 5-6. Sample Change Order Register

trade contractors and writing clear contract documents. Use facts, not emotions, when preparing to terminate a trade contract.

In most cases, everyone loses in both the short and long term when a contract needs to be terminated. Termination is a costly event. Not just time and money, but stress and reputation are affected. Have a back up contractor available and ready to take over the work. Consult the evaluations of the trade contractors who responded to your request for a quote and contract with the next best contractor.

If early termination is necessary, a well documented file is essential in the event of litigation. The final notice should be brief and to the point. Once decided, carry out the termination promptly and bring the replacement contractor on board. Any discussions with the contractor regarding the termination should be witnessed.

Communicate with Contractors

Communication is crucial to effective management. Builders need to be able to impart information and changes to trade contractors effectively and in a timely manner.

Communication also is an effective tool for team building. Open communication lines between builders and trade contractors can help ensure that the project proceeds on schedule and according to quality controls.

Hold Preconstruction Meetings

On large or complex projects, schedule a preconstruction meeting with all trade contractors before starting work. The purpose of the meeting is to:

- introduce all members of the construction team
- obtain their input on the construction schedule and the sequencing of trade contractor work
- discuss jobsite management issues such as safety, cleanup, and coordination

A sample agenda for a preconstruction meeting is shown in Figure 5-7. This meeting is crucial to building a good foundation for timely construction of the project. It is important that the builder's superintendent establish a good rapport with the trade contractors at this meeting. The success of the project depends on these relationships.

Arrange Periodic Meetings

Most communications with trade contractors are by telephone or at informal jobsite meetings. Maintaining frequent communications with the trades is essential to fostering a team relationship and ensuring a quality project that is completed on time.

PRECONSTRUCTION MEETING AGENDA

1. Introduction of Key Personnel

   - Builder's Project Staff

   - Trade Contractors' Project Staffs

2. Discussion of Project Schedule

3. Physical Organization of Project Site

   - Identification of Areas for Contractors' Storage of Materials

   - Identification of Parking Areas for Contractors' Crews

4. Site Safety Requirements and Responsibilities

5. Contract Change Order Procedures

6. Inspection of Completed Contractor Work

7. Site Cleanup Requirements and Responsibilities

8. Contractor Payment Procedures

9. Warranty Procedures

Figure 5-7. Sample Preconstruction Meeting Agenda

Conducting scheduled weekly meetings with trade contractors allows the entire project team to focus on issues such as schedule changes. Such meetings support the builder's team-building efforts and can have a positive impact on project execution.

Request Information

If the contract drawings lack sufficient detail for a trade to understand what is expected, the contractor needs a simple way to ask for clarifica-

Cascade Homes

1650 Happy Valley Road
Olympia, Washington 98507

REQUEST FOR INFORMATION

Project: _LOT 30, HIDDEN COVE_ Date: _DEC. 3, 1997_

Contractor: _JAY'S PLUMBING_

RFI Number: _1_

To: _BILL WALLACE, CASCADE SUPERINTENDENT_

Description of Request:

PLEASE PROVIDE INFORMATION FOR WHIRLPOOL TUB. DO YOU WANT COLOR TO MATCH OTHER FIXTURES? PLEASE SPECIFY MODEL NUMBER.

Response Needed By: _DEC. 10, 1997_

From: _Ken Adams_

Response:

PRICE THE BONE COLOR. MODEL NUMBER IS 36040.

From: _Bill Wallace_ Date: _DEC. 8, 1997_

Figure 5-8. Sample Request for Information Form

tion. Forms, such as the one shown in Figure 5-8, should be made available to contractors so they can request the needed information. Request forms should be easy to use and provide a space for a response.

As a builder, you should maintain a log of all requests for information received from trade contractors, including the date of receipt and the status of the response. This ensures prompt responses are provided and the trade contractors' progress is not impeded. A sample request for information log is shown in Figure 5-9. Blank copies of the request for information form and the information log are included in the Appendix and on the diskette.

Maintain a Document Register

Builders should maintain a log of any documentation required from trade contractors. Include the time of receipt and the status of approval in the log. Examples of required documentation include samples of carpet, wallpaper, cabinet material, paint colors, bathroom tile, and other finish items. Timely submission and approval of all required documentation helps maintain the integrity of the project schedule. A sample

---

Cascade Homes
1650 Happy Valley Road
Olympia, Washington 98507

REQUEST FOR INFORMATION LOG

Project: _LOT 30, HIDDEN COVE_

RFI No.	Submitted By	Subject	Date Rec'd	Date Ans'd	Remarks
1	JAY'S PLUMB.	WHIRLPOOL TUB	12/3/97	12/8/97	CHANGE ORDER
2	PARKER CARP.	CEILING GARAGE	1/6/98	1/10/98	
3	FRANK ELEC.	CEILING FANS	1/15/98	4/20/98	

Figure 5-9. Sample Request for Information Log

document register is shown in Figure 5-10. A blank copy is included in the Appendix and on the diskette.

Keep Job Diaries

Record the activities of all trade contractors who work on the project each day in a daily job diary. Include information such as work done, number of workers, types of equipment, and quantity of materials received. Weather conditions and conversations are also items that should be recorded. The objective of the diary is to maintain a daily record of all activities occurring on the project. Although quality trade contrac-

---

Cascade Homes

1650 Happy Valley Road
Olympia, Washington 98507

DOCUMENT CONTROL REGISTER

Project: LOT 30, HIDDEN COVE

Contractor: JAY'S PLUMBING

Description	Date Received	Date Approved	Remarks
COLOR SAMPLES	11/1/97	11/5/97	BONE
MANUFACTURER'S SPEC FOR WHIRLPOOL TUB	12/15/97	12/20/97	APPROVED

Figure 5-10. Sample Document Control Register

tors seldom cause serious problems, this information can be crucial when addressing requests for change orders or back charges by trade contractors.

You also can use the job diary to record safety or disciplinary problems, including notes about appropriate follow-up as needed to protect yourself from potential liability. Such a record adds to the base of historical information you can use when planning for future projects. The diary should be handwritten and maintained in a bound book to help ensure the legitimacy of the entries should the diary ever be required to serve as evidence in a legal matter and to lessen the likelihood that specific pages or sections of the diary might be lost. If there is ever a serious injury, fatality, or OSHA citation related to a contractor with past safety problems (as indicated in the diary), consult with your labor attorney immediately about the existence of the information.

Typed or word-processed diary entries are susceptible to tampering and can be made out of sequence. The identity of the person making the entries also may be questioned. Entries are likely to be more consistent and accurate if made on the spot at the jobsite rather than later back at the office. A sample entry in a daily diary is shown in Figure 5-11. A blank copy is included in the Appendix and on the diskette.

Establish Safety Controls

The OSHA standards apply to all employees including builders and their trade contractors. Builders are fully responsible for all safety concerns on their jobsites. They cannot delegate this responsibility to the trades. Trade contractors are responsible for the safety of their employees. All trade work must be done in compliance with OSHA standards.

It is good practice to include your safety requirements for the trades in their contracts and hold them liable for the safety of their work crews. A complete safety program is beyond the scope of this book, but some minimum aspects of safety control include a safety plan with the following:

- A statement of the builder's commitment to safety
- Recognition of the safety concerns of the workers
- Recognition that 75 percent of construction accidents come from worker behavior and 25 percent from the work environment
- The requirement that each trade comply with safety standards
- Identification of hazards on the jobsite
- Identification of methods for removing and avoiding hazards or devices that protect workers from hazards
- A requirement that Material Specification Data Sheets (MSDS) for hazardous materials be acquired and maintained on the jobsite
- Requirements for safety training, inspections, and documentation

Cascade Homes

1650 Happy Valley Road
Olympia, Washington 98507

DAILY PROJECT REPORT

Project: _LOT 30, HIDDEN COVE_   Date: _DEC. 15, 1997_

Superintendent: _Bill Wallace_

Weather: _LIGHT SHOWERS_

Temperature: a.m. _62°_ p.m. _70°_

Visitors: _____

POWER COMPANY REP.
NELSON PAINT - BOB JAMES

Description of Work

FINISHED ROOF SHEATHING.
ROUGH IN PLUMBING.
STARTED ELECTRICAL
WORK.

Personnel

Number	Trade
6	Carpentry
	Drywall
1	Electrical
	Fire Protection
	Flooring
	Glazing
2	Heating & Air Conditioning
	Insulation
	Landscaping
	Masonry
	Metal Framing
	Millwork
	Painting
2	Plumbing
	Siding
	Sitework
	Stucco
11	Total

Materials Received

ROOF SHINGLES - 95 BNDL
FELT - 10 ROLLS

Figure 5-11. Sample Daily Project Report

In addition to their own safety program, builders should require each trade to furnish their job-specific safety plan. The plans do not need to be lengthy, but should address any identifiable job hazards under control of the trade and a method to prevent employees from being subject to the hazard. Figure 5-12 presents an excerpt from a construction safety checklist developed for the NAHB Labor, Safety, and Health Department's *Model Safety and Health Program for the Building Industry*. The complete checklist is reproduced in Appendix 1.

Conduct Post-Construction Surveys

Some builders find it useful to conduct a post-job survey of all trade contractors involved in a project. The survey requests helpful feedback on all aspects of the contractor's participation in the project. Trade contractors who feel comfortable providing such feedback, whether verbally or in a written survey, often can contribute cost- or time-saving ideas the builder can use when planning future projects.

Require Coordinated Warranty Service

Clearly communicate to trade contractors the procedures to be followed in the event the homeowner requests warranty service. Some builders allow homeowners to contact trade contractors directly, while others re-

---

CONSTRUCTION SAFETY AUDIT CHECKLIST
Conditions to Check

1. **Housekeeping and Sanitation:**
   - ☐ General condition of work areas
   - ☐ Adequate trash removal
   - ☐ Floor openings covered or guarded
   - ☐ Stairs and walkways cleared of debris and materials
   - ☐ Note any slip, trip, or fall hazards; guardrails erected on stairways, wall openings, etc.
   - ☐ Adequate lighting
   - ☐ Adequate ventilation
   - ☐ Toilet facilities adequate
   - ☐ Drinking water and cups provided

2. **Personal Protective Equipment Issued and Used as Instructed:**
   - ☐ Hard hats
   - ☐ Protective glasses and goggles
   - ☐ Gloves
   - ☐ Respirators

3. **Ladders and Scaffolding:**
   - ☐ In good, serviceable condition
   - ☐ Properly positioned and secured at the top
   - ☐ Extend 36" above roof or platform, if used for access

Figure 5-12. Excerpt from Construction Safety Checklist

Created from Exhibit 6-3 of the *Model Safety and Health Program for the Building Industry*, 2d ed. (NAHB Labor, Safety, and Health Services Department, 1995), 24–25.

quire all warranty service requests to be processed through them. In some locations, builders may be able to contract with a customer service company to manage warranty service requests. Select the method that best fits your management style. What is crucial is that both homeowners and trade contractors understand the procedures and their respective responsibilities. Responsive warranty service procedures are essential to ensuring homeowner satisfaction.

Project Success

Effectively managed trade contracts are essential to successful residential construction. The work requires that you clearly establish and communicate with the trades the specific scope of work they are to perform. Once trade contracts are developed, select the best team of trade contractors for each project using the evaluation criteria. After the trade contractors are selected, incorporate them into your construction team. Use management strategies discussed in this chapter to build long-term relationships with quality trade contractors and you will find success in your building projects.

Appendix 1

Blank Forms

The forms that appear on the following pages have been adapted from the figures used in this book. These documents and the model scopes of work from Chapter 3 have also been reproduced as files on the diskette that accompanies this book.

ATTENTION: The model language provided in the sample forms and on the diskette is provided as a convenience to builders who wish to incorporate applicable language into their own contract documents. This material should not be used without the review and approval of an attorney experienced in construction contract law. Builders adapting this material for use in their businesses should have their attorneys prepare specific documents that will meet their particular needs.

[Builder's Company Name]

[Builder's Business Address]

CONSTRUCTION CONTRACT

THIS AGREEMENT, made and entered into this _____ day of _____, 19__, by and between [Builder's Company Name] and [Trade Contractor's Company Name], herein called the "Subcontractor," for all [Trade Specific] work associated with the construction of the [Owner's Name] residence located at [Jobsite Address].

WHEREAS, [Builder's Company Name] entered into a contract dated the _____ day of _____ 19__ with [Owner's Name], herein called the "Owner," for the construction of a residence according to the terms and conditions of said contract and the general specifications and supplements, addenda, general and special conditions, plans, drawings, and other documents made a part thereof, and any change orders or amendments, collectively referred to as the "general contract."

WHEREAS, Subcontractor acknowledges that it is familiar with the general contract and agrees that the general contract is a part hereof and incorporated as a part of this subcontract.

IT IS HEREBY AGREED AS FOLLOWS:

WORK TO BE PERFORMED. Subcontractor agrees to provide all materials, labor, tools, equipment, supervision, supplies, and other items necessary or required to execute the work and perform all [Trade Specific] work in accordance with the contract drawings, specifications, and any addenda contained in the general contract documents for the construction of the [Owner's Name] residence. The intent is to provide a complete [Trade Specific] system that meets local code requirements including, but not limited to, provision and installation of the following:

1. _____

2. _____

3. _____

4. _____

5. _____

6. _____

Continued

Trade Contract, Page 2

CONTRACT PRICE. [Builder's Company Name] agrees to pay and Subcontractor agrees to accept as full compensation for performing all work according to the requirements of the contract and furnishing all materials, supplies, and equipment required to execute the work, the sum of [Amount Spelled Out] *dollars* ($[Amount as Figure]) subject to additions and deductions from such changes in the scope of work as agreed upon in writing through change orders to the contract.

TERMS AND CONDITIONS OF THIS CONTRACT ARE ATTACHED.

Builder: [Builder's Company Name]

By: ➤ _____
                              Authorized Signature/Title

Date: _____/_____/_____

Subcontractor: [Trade Contractor's Company Name]

By: ➤ _____
                              Authorized Signature/Title

Date: _____/_____/_____

Continued

TERMS AND CONDITIONS OF CONSTRUCTION CONTRACT

Between

[Builder's Company Name] and [Trade Contractor's Company Name]

Dated [Date]

1. GENERAL RESPONSIBILITIES

a. Subcontractor shall comply with the General Conditions of the general contract for the construction of the [Owner's Name] residence and any interpretations as to the meaning thereof issued by the Owner.

b. Subcontractor shall ensure that all wiring and electrical items fully comply with code requirements that are effective on the date contract is signed. Code modifications made after contract has been signed are not included.

c. Subcontractor shall visit the construction site before starting work to understand access restrictions to the site.

d. Subcontractor shall coordinate the installation of all [Trade Specific] work with other interfacing trades to ensure an unencumbered worksite.

e. Subcontractor shall coordinate connection with the local [Trade Specific] utility.

f. Subcontractor shall [Additional Trade Specific Coordination].

g. Subcontractor shall provide a qualified onsite supervisor whenever work is being performed. Supervisor will attend weekly coordination meetings to review scheduled versus actual progress, quality control, coordination of work, and jobsite safety.

h. Subcontractor shall submit required shop drawings and manufacturers' cut sheets for all equipment and await receipt of approval of Owner or Architect prior to starting work.

i. Subcontractor shall at all times keep the project site clean of dirt, debris, trash, and any waste materials arising from the performance of the subcontract. Subcontractor is responsible for removal of all debris created as a result of the work being performed and disposal at a site designated by [Builder's Company Name].

2. COMMENCEMENT AND PROGRESS OF WORK

a. Subcontractor agrees to comply with and perform the subcontracted work in conformance with project plans and specifications and applicable state, county, and municipal codes, ordinances, and statutes according to the requirements of the construction schedule or schedules as [Builder's Company Name] may from time to time develop and submit to Subcontractor. Within three (3) calendar days after being notified by [Builder's Company Name], Subcontractor shall commence actual construction on such parts of the scope of work as [Builder's Company Name] may designate and to thereafter continue diligently in the performance of the work. All work shall be performed in full cooperation with [Builder's Company Name] and other subcontractors.

b. Upon request, Subcontractor shall prepare and submit to [Builder's Company Name] for approval a progress schedule to meet the dates as shown by [Builder's Company Name]'s then-current construction schedule and showing the order in which Subcontractor proposes to carry on the work and the dates on which it will start and complete salient features of the subcontracted work.

Continued

c. If in the opinion of [Builder's Company Name] Subcontractor falls behind the progress schedule, Subcontractor shall take steps as may be necessary to improve the subcontract progress, and [Builder's Company Name] may require Subcontractor to increase the number of shifts and/or overtime work, days of work, and/or increase equipment and/or tools being used, and to submit such revised schedule demonstrating the manner in which the agreed rate of progress will be regained.

3. PAYMENT PROCEDURES

a. Payment will be made to Subcontractor for work performed under this subcontract as measured and certified as being completed by [Builder's Company Name]. Before any payment is due, Subcontractor must provide [Builder's Company Name] with a written estimate of the total amount of work completed and a signed statement that all completed work fully conforms to contract requirements.

b. Upon application, partial payments for work performed under this subcontract will be made by [Builder's Company Name], and will equal the value of the work performed by Subcontractor, less the sum of previous payments.

c. Subcontractor will receive partial payments fifteen (15) days after invoice has been received by [Builder's Company Name].

d. Upon completion of all contracted work, Subcontractor will be paid the remaining amount due Subcontractor under this subcontract. Final payment shall release [Builder's Company Name] from any further obligations whatsoever in respect to this subcontract. Subcontractor shall, as a condition to receipt of final payment, provide to [Builder's Company Name] a full release from any and all claims, liens, and demands whatsoever for all matters growing out of, or in any matter connected with this subcontract.

4. LAWS AND REGULATIONS

Subcontractor, its employees and representatives, shall at all times comply with all applicable laws, ordinances, statutes, rules, and regulations, federal and state, county and municipal, and particularly those relating to wages, hours, fair employment practices, nondiscrimination, and working conditions. Subcontractor shall procure and pay for all permits, licenses, and inspections required by any governmental authority for any part of the work under this subcontract and shall furnish any bonds, security, or deposits required by such authority to permit performance of the work.

5. INSURANCE

a. Subcontractor, at its own expense, shall procure, carry, and maintain on all of its operations workers' compensation and employer's liability insurance covering all of its employees, public liability and property damage insurance, and automotive public liability and property damage insurance. Coverage limits shall be in accordance with the requirements of the general contract. Subcontractor is required to name [Builder's Company Name] and Owner as additional insureds on Subcontractor's general liability policy.

b. Subcontractor shall provide to [Builder's Company Name] prior to commencement of work a certificate from the insurance companies that such insurance is in force and will not be canceled without thirty (30) days written notice to [Builder's Company Name].

Continued

6. INDEMNIFICATION

Subcontractor shall indemnify and hold harmless [Builder's Company Name] and Owner against any claims, damages, losses, and expenses, including legal fees, arising out of or resulting from performance of subcontracted work to the extent caused in whole or in part by the Subcontractor or anyone directly or indirectly employed by the Subcontractor.

7. SAFETY

a. Subcontractor and all of its employees shall follow all applicable safety and health laws and requirements pertaining to its work and the conduct thereof, but not limited to, compliance with all applicable laws, ordinances, rules, regulations, and orders issued by a public authority, whether federal, state, or local, including the Federal Occupational Safety and Health Administration, and any safety measures required by [Builder's Company Name] or Owner.

b. Safety of Subcontractor's employees, whether or not in common work areas, is the responsibility of Subcontractor.

c. Subcontractor agrees to instruct all its employees to inform [Builder's Company Name] immediately of any unsafe condition or practice whether or not in common work areas.

8. INSPECTION AND ACCEPTANCE OF COMPLETED WORK

The materials and work shall at all times be subject to inspection by Owner and [Builder's Company Name]. Owner and [Builder's Company Name] shall be afforded full and free access to the construction site for the purpose of inspection and to determine the general progress of the work. In the event that any of the work or materials are found to be improper or defective by Owner or [Builder's Company Name], Subcontractor shall upon notification in writing from [Builder's Company Name] proceed to replace or correct the defective material or workmanship at its own cost and expense. If Subcontractor fails to correct the defective work, [Builder's Company Name], at its option, may replace and correct the same and deduct the cost of correcting the defective work from Subcontractor's payments.

9. CHANGE ORDER PROCEDURES

[Builder's Company Name] may order additional work, and Subcontractor will perform such changes in the work as directed in writing. Any change or adjustment to the subcontract price as a result of changes in the scope of work shall be as specifically stated in the change order. If Subcontractor encounters conditions it considers different from those described in subcontract documents or plans, it is required to issue written notice to [Builder's Company Name] before proceeding. Subcontractor's failure to issue notice shall constitute waiver of any claims for additional compensation. Only [Builder's Company Name]'s project superintendent is authorized to issue change orders to Subcontractor. If Subcontractor and [Builder's Company Name] cannot agree upon a price for the changes in the work, [Builder's Company Name] may direct Subcontractor to execute the changes, and Subcontractor will be paid based on the actual cost to Subcontractor plus a reasonable markup for profit and overhead expenses.

Continued

10. DISPUTE RESOLUTION

Any dispute between the Subcontractor and [Builder's Company Name] arising out of or related to this subcontract that is not informally resolved shall be settled by arbitration under the procedures contained in the general contact. Any Subcontractor claims for additional compensation or damages due to acts or omissions of Owner shall be submitted to [Builder's Company Name], who will submit the claim to the Owner on behalf of Subcontractor. Resolution procedures will be those contained in the general contract.

11. TRADE CONTRACTOR'S STATUS AS INDEPENDENT CONTRACTOR

The relationship between [Builder's Company Name] and Subcontractor is that of independent contractor. Subcontractor has the status of an employer as defined by the Unemployment Compensation Act of the state in which this contract is to be performed and all similar acts of the national government including all Social Security Acts. Subcontractor will withhold from its payrolls as required by law or government regulation and shall have full and exclusive liability for the payment of any and all taxes and contributions for unemployment insurance, workers' compensation, and retirement benefits that may be required by federal or state governments.

12. WARRANTY

   a. Subcontractor warrants its work under this subcontract against all deficiencies and defects in material and/or workmanship. All materials and equipment furnished shall be new and installed in conformance with code requirements. Subcontractor agrees to repair or replace at its expense and pay for any damages resulting from any defective materials or workmanship that appear within one (1) year after the closing date.
   b. Subcontractor shall remedy, at its expense, any defects due to faulty materials or workmanship and pay for any damages caused to other work within seven (7) working days after being notified by [Builder's Company Name] or Owner.
   c. [Builder's Company Name] retains the right to perform remedial work not completed within the established time frame and charge the cost to Subcontractor.

13. TERMINATION OR SUSPENSION

   a. If Subcontractor fails to carry out the work in accordance with this subcontract and fails within seven (7) days after receiving written notice to make correction, [Builder's Company Name] will issue a second letter of noncompliance and notify Subcontractor of its intent to terminate the subcontract. If Subcontractor fails to make the necessary corrections within three (3) working days after receipt of the second letter, [Builder's Company Name] may terminate the subcontract.
   b. [Builder's Company Name] may, without cause, order Subcontractor in writing to suspend, delay, or interrupt the work in whole or in part for a period of time. An adjustment in subcontract price shall be made to cover the cost of performance, including profit on the increased cost, caused by the suspension, delay, or interruption.

[Builder's Company Name]

[Builder's Business Address]

TRADE CONTRACTOR QUESTIONNAIRE

1. General Company Information

Name:_____

Address:_____

City:_____ State:_____ Zip:_____

Telephone:_____ FAX:_____

(Check appropriate box.)

❒ Sole owner. Name:_____

❒ Partnership. List names of partners:

_____

_____

_____

_____

❒ Corporation.

President:_____

Vice President:_____

Number of years in business under present name:_____

Continued

1. **General Company Information** (continued)

Trade(s) normally performed by company: _____

_____

Number of trade and office personnel currently employed:

Trade employees: _____   Office employees: _____

2. **Financial Information**

What is the maximum dollar value of work the company is capable of handling at one time?

$ _____

Attach last 2 years audited financial statements at end of questionnaire.

3. **Insurance Information**

What is the company's workers' compensation experience modification rate for the 3 most recent years?   19____ _____   19____ _____   19____ _____

How much insurance coverage does the company currently carry?

	Yes	No	Amount
General Liability	___	___	_____
Automobile Liability	___	___	_____
Workers' Compensation	___	___	_____

4. **Safety Information**

Does the company have a written safety program?   _____ Yes   _____ No

Use your OSHA Form 200 to complete the following table.

	19____	19____	19____
Total Number of Workers' Compensation Claims			
Number of Lost Time Workers' Comp. Claims			
Number of Accident Liability Claims			
Number of Fatalities			

Continued

5. Project Information

List current, ongoing projects with approximate dollar value and estimated completion date.

Project	Amount	Completion Date

Has the company failed to complete any work assigned to it during the past 5 years?

_____ Yes   _____ No

If yes, explain:

_____

_____

_____

Attach a list of projects completed in the last 3 years and a list of builder references whom we may contact.

6. Equipment Information

Attach a list of owned construction equipment with capacity, age, type, and attachments.

This questionnaire was completed by:

Name:_____ Title:_____

Signature:_____ Date:_____

[Builder's Company Name]

[Builder's Business Address]

REQUEST FOR QUOTATION

To: [Trade Contractor's Company Name]   RFQ #: _____
      [Trade Contractor's Business Address]   Date: _____

Project: [Owner's Name] Residence

Project Address: [Jobsite Address]

Quote Required by: [Date Required]

Contract Documents: General contract for construction of [Owner's Name] Residence prepared by [Builder's or Architect's Company Name].

Scope of Work: Provide all materials, labor, tools, equipment, supervision, supplies, and other items necessary or required to safely execute the work and perform all [Trade Specific] work in accordance with the contract drawings, specifications, and any addenda contained in the general contract documents for the construction of the [Owner's Name] residence. The intent is to provide a [Trade Specific] system that meets local code requirements including, but not limited to, provision and installation of the following:

1. _____
2. _____
3. _____
4. _____
5. _____
6. _____

Contractor's quote will be based on the complete set of contract documents that are available at [Builder's Company Name]'s business office.

Terms and conditions of the [Trade Specific] contract are attached.

Continued

To: [Builder's Company Name]
      [Builder's Business Address]

We propose to perform the entire scope of work for [_____ Dollar Amount _____].

We agree to perform the described work for the quoted price if notified of acceptance by [Builder's Company Name] within 60 days of date submitted.

Subcontractor: [Trade Contractor's Company Name]

By: ➤ _____
                        Authorized Signature/Title

Date: _____/_____/_____

[Builder's Company Name]

[Builder's Business Address]

QUOTATION EVALUATION SHEET

Project:_____

Contract Scope:_____

	COST
Estimated Cost: [by Individual's Name]	
CONTRACTOR	

[Builder's Company Name]

[Builder's Business Address]

PROJECT INSPECTION REPORT

Project: _____ Date: _____

Inspection Participants:

_____

_____

_____

_____

_____

_____

Scope of Inspection:

_____

_____

_____

Items Requiring Correction:

_____

_____

_____

_____

_____

_____

_____

_____

_____

_____

_____

_____

PARTIAL LIEN RELEASE
UPON RECEIPT OF PROGRESS PAYMENT

Upon receipt by the undersigned of a check from [Builder's Company Name   ]

in the sum of $ [Amount] payable to [Trade Contractor's Company Name   ]
and when the check has been properly endorsed and paid by the bank upon which it is drawn,
this document shall become effective to release any mechanic's lien rights the undersigned has
on the project of:

[Lot Number, Community Name   ] located at [City, State   ]
to the following extent. This release covers a progress payment for labor, services, material,
and equipment furnished to:

[Builder's Company Name   ] through [Date Covered by Progress Payment] only
and does not cover labor, services, material, or equipment not compensated by the progress
payment.

Date: _____   [Trade Contractor's Company Name   ]

                                             By: [Authorized Signature   ]

FINAL LIEN RELEASE
UPON RECEIPT OF FINAL PAYMENT

Upon receipt by the undersigned of a check from [Builder's Company Name   ]

in the sum of $ [Amount   ] payable to [Trade Contractor's Company Name   ]
and when the check has been properly endorsed and paid by the bank upon which it is drawn,
this document shall become effective to release any mechanic's lien rights the undersigned has
on the project of:

[Lot Number, Community Name   ] located at [City, State   ].
The undersigned has been paid in full for all labor, services, material and equipment
furnished to:

[Builder's Company Name   ].

Date: _____   [Trade Contractor's Company Name   ]

                                         By: [Authorized Signature   ]

[Builder's Company Name]

[Builder's Business Address]

CONTRACT CHANGE ORDER

Change Order Number: _____

Project: _____

Contract: _____

We agree to make the following change(s) in the contract scope of work:

_____

_____

_____

Change(s) will affect the following plans and/or specifications:

_____

Subject to the following adjustment to the contract value:

     Cost of this Change: _____

     Previous Contract Amount: _____

     Revised Contract Amount: _____

     Revised Completion Date: _____

Builder: [Builder's Company Name]

By: ➤_____ Date: _____
            Authorized Signature/Title

Subcontractor: [Trade Contractor's Company Name]

By: ➤_____ Date: _____
            Authorized Signature/Title

[Builder's Company Name]

[Builder's Business Address]

CHANGE ORDER REGISTER

Project: _____

Trade Contractor: _____

Date Awarded: _____

				Contract Value
		Original Contract Value		
Change Order Number	Date of Change Order	Scope of Change Order	Change Order Value	

[Builder's Company Name]

[Builder's Business Address]

REQUEST FOR INFORMATION

Project: _____ Date: _____

Contractor: _____

RFI Number: _____

To: _____

Description of Request:

Response Needed By:

From:

Response:

From: Date:

[Builder's Company Name]

[Builder's Business Address]

REQUEST FOR INFORMATION LOG

Project: _____

RFI No.	Submitted By	Subject	Date Received	Date Answered	Remarks

[Builder's Company Name]

[Builder's Business Address]

DOCUMENT CONTROL REGISTER

Project: _____

Contractor: _____

Description	Date Received	Date Approved	Remarks

[Builder's Company Name]

[Builder's Business Address]

DAILY PROJECT REPORT

Project: _____ Date: _____

Superintendent: _____

Weather: _____

Temperature: a.m. _____ p.m. _____

Visitors: _____

_____

_____

_____

Personnel

Number	Trade
	Carpentry
	Drywall
	Electrical
	Fire Protection
	Flooring
	Glazing
	Heating & Air Conditioning
	Insulation
	Landscaping
	Masonry
	Metal Framing
	Millwork
	Painting
	Plumbing
	Siding
	Sitework
	Stucco
	Total

Description of Work

_____

_____

_____

_____

_____

_____

_____

_____

_____

_____

_____

_____

_____

Materials Received

_____

_____

_____

_____

_____

_____

_____

_____

_____

_____

CONSTRUCTION SAFETY AUDIT CHECKLIST

Conditions to Check

1. **Housekeeping and Sanitation:**
   - ☐ General condition of work areas
   - ☐ Adequate trash removal
   - ☐ Floor openings covered or guarded
   - ☐ Stairs and walkways cleared of debris and materials
   - ☐ Note any slip, trip, or fall hazards; guardrails erected on stairways, wall openings, etc.
   - ☐ Adequate lighting
   - ☐ Adequate ventilation
   - ☐ Toilet facilities adequate
   - ☐ Drinking water and cups provided

2. **Personal Protective Equipment Issued and Used as Instructed:**
   - ☐ Hard hats
   - ☐ Protective glasses and goggles
   - ☐ Gloves
   - ☐ Respirators

3. **Ladders and Scaffolding:**
   - ☐ In good, serviceable condition
   - ☐ Properly positioned and secured at the top
   - ☐ Extend 36" above roof or platform, if used for access
   - ☐ Doors blocked open, locked, or guarded off if in front of ladder
   - ☐ Stepladders fully open when used
   - ☐ Metal ladders not used for work in electrical areas
   - ☐ Sound, rigid footing for all scaffolds
   - ☐ Safe access to all working levels
   - ☐ Equipped with standard guardrails, midrails, and toeboards
   - ☐ Protection provided where persons are required to work or pass under scaffolding in use

   - ☐ No accumulation of tools or material on platforms
   - ☐ Outriggers installed, if required
   - ☐ Self-propelled (motorized) types of scaffolds require special maintenance and inspection

4. **Portable Power and Hand Tools:**
   - ☐ General condition of tools
   - ☐ Proper tool being used for job being performed
   - ☐ Guards and safety devices are operable and in place
   - ☐ Electrical tools inspected and marked according to the Assured Equipment Grounding Program
   - ☐ Tool retainers used on pneumatic tools; air pressure properly regulated
   - ☐ Check for pinch and shear points

5. **Powder-Actuated Tools:**
   - ☐ All operators trained and certified
   - ☐ Tools and charges protected from unauthorized use
   - ☐ Loaded tools are not left unattended
   - ☐ All tools inspected and tested daily before use
   - ☐ Tools and charges matched to recommended materials only
   - ☐ Safety goggles or face shields used by operators
   - ☐ Local regulations complied with

6. **Welding and Cutting:**
   - ☐ Operators trained and qualified
   - ☐ Personal protective equipment
   - ☐ Fire extinguishers provided
   - ☐ Flammable materials protected

Continued

- ❐ Gas cylinders secured
- ❐ All fittings free of oil and grease
- ❐ Flashback protection used
- ❐ Proper gauge settings
- ❐ All hoses, cords, and other equipment in good condition

7. All Material Storage and Handling:
- ❐ Materials properly stacked and on firm footings; properly blocked and secured
- ❐ Fire protection adequate
- ❐ All rigging and lifting equipment properly maintained and inspected periodically
- ❐ Employees picking up and handling loads properly
- ❐ Flammable liquids stored only in approved containers
- ❐ Flammable gases properly stored
- ❐ Adequate security measures

8. Fall Protection:
- ❐ Guardrails provided where necessary
- ❐ Personal Fall Arrest Systems (PFAS) provided where necessary
- ❐ Fall protection plan implemented where PFAS not used

9. Excavation and Trenching:
- ❐ All excavations or trenches properly shored
- ❐ All excavations have proper means of entry and exit
- ❐ Excavations or trenches inspected by competent person

10. First Aid Kits:
- ❐ Kits provided
- ❐ Kits inspected and replenished where necessary

Project Name: _____

Project Number: _____

Superintendent: _____

Date: _____

Created from Exhibit 6-3 of the *Model Safety and Health Program for the Building Industry,* 2d ed. (NAHB Labor, Safety, and Health Services Department, 1995), 24-25.

Diskette Information

The files on the diskette that accompanies this book are word processed files. They are *not* program software; **do not attempt to boot from this diskette.** The diskette contains files for each of the sample forms that appear in Appendix 1 and the sample scopes of work for all 33 trades that appear in Chapter 3.

All of the files on this diskette appear in three formats: Microsoft® Word for Windows™ (Word 6.0/95), WordPerfect 5.1/5.2 for Windows™, and MS DOS® Text (ASCII text). These three formats have been provided in order to make the files available to a larger number of builders. The Word files appear on the diskette in a folder, or directory, named **wrdscope**; the WordPerfect files are in a directory named **wpscope**; and the MS DOS Text files are in a directory named **dosscope**.

The forms reproduced in Word and WordPerfect from Appendix 1 include simple formatting (bolding, fonts, page layout codes) so that you may print and use the forms with minimal adaptation. However, the model scopes of work from Chapter 3 are intended to be modified and incorporated into builders' existing contract documents. Therefore, the files containing the model scopes of work have been provided *without* page formatting so that builders may more easily adapt them to their own document styles. The MS DOS Text (ASCII text) files do not contain any page formatting.

Using the Diskette

You may open the word processed files directly from your computer's diskette drive, but we recommend copying the files to your computer's hard drive and retaining the diskette as a backup. To put a working copy on your computer's hard drive, insert the diskette into your 3.5-inch drive and follow these instructions.

MS-DOS Users (version 6.0 or later)

At the DOS prompt, type **a:install** and press **Enter;** then follow the prompts to select the ASCII version of the software. DOS will copy the appropriate files onto your hard drive.

Windows 3.1 Users

In Program Manager, go to the **File** menu and select **Run.** A dialog box will appear. On the command line, type **a:install** and click on **OK.** Follow the prompts to select the Word, WordPerfect, or DOS version of the software. Windows will copy the appropriate files onto your hard drive.

Windows 95 Users

Click on the **Start** icon and select **Run.** A dialog box will appear. On the **Open** line, **A\:install.bat** should appear. Click on **OK.** Follow the prompts to select the Word, WordPerfect, or DOS version of the software. Windows will copy the appropriate files onto your hard drive. After all the files have been copied, click on the **x** in the right hand corner of the window.

If you install the diskette files using the above procedures, the files will be placed in a directory called **c:\scopes** on your computer's hard drive. (The files will appear in this directory under the subdirectory **wrdscope, wpscope,** or **dosscope,** depending on which version you have chosen to install.) If you already have file directories for documents relating to contracts and scopes of work you may wish to move the files from **c:\scopes** into your preferred directory or subdirectory. To do so, follow the instructions for **Move** or **Copy** in your DOS or Windows manual. **Caution:** If you already have a directory called **c:\scopes** on your computer's hard drive you may need to rename this directory before running this install program.

Note: If your system designates the 3.5-inch drive as the "b" drive, substitute **b:install** for **a:install** in the instructions, and select the appropriate option to copy files "from B."

Other Software

Your word processing software may be able to read Word or Word-Perfect files. Check your software manual and follow the directions for opening and converting files from Word. You may have to clean up or recreate some of the page formatting.

If your software is unable to read or convert the Word or WordPerfect files, run the installation program and select the files that have been saved as MS DOS Text (ASCII text) files. Remember, page formatting is not a part of these files; they will contain all the text content but you will have to reconstruct the layouts.

Printers vary in how they interpret font and layout codes from word processing software. In most cases, only minor cleanup or reformatting will be required. Some difficulties can be resolved by checking to be sure you are using the printer *driver* supplied by your word processing software; printer drivers that come with word processing software generally support a wider range of fonts than the printer driver that comes with the printer itself.

Remember to treat the files on this diskette as you would any other word processed files. Before using them, first open Word or whatever word processing software you use. Be sure to make a complete set of backup copies and save working files regularly to prevent the accidental loss of formatting or data.

Should you ever need to reprint these instructions this page has been saved on the diskette in a file named **readme.**

Filenames

Identical filenames have been used for the Word, WordPerfect, and MS DOS Text (ASCII text) versions of these files. Filenames for Word files use the extension **.doc**; filenames for WordPerfect files use the extension **.wpd**; and filenames for MS DOS Text (ASCII text) files use the extension **.txt.**

			File Extension in	
Sample Form or Model Scope	Filename	Word	WordPerfect	ASCII
Trade Contract	CONTRACT	.doc	.wpd	.txt
Trade Contractor Questionnaire	TQQUEST	.doc	.wpd	.txt
Request for Quotation	QUOTEFRM	.doc	.wpd	.txt
Quotation Evaluation Sheet	QUOTEVAL	.doc	.wpd	.txt
Project Inspection Report	INSPECT	.doc	.wpd	.txt
Partial Lien Release	RELEASE	.doc	.wpd	.txt
Final Lien Release	RELEASE2	.doc	.wpd	.txt
Contract Change Order	CHANGE	.doc	.wpd	.txt
Change Order Register	CXREGIST	.doc	.wpd	.txt
Request for Information	INFOREQ	.doc	.wpd	.txt
Request for Information Log	INFOLOG	.doc	.wpd	.txt
Document Control Register	DOCCTRL	.doc	.wpd	.txt
Daily Project Report	DAILY	.doc	.wpd	.txt
Safety Audit Checklist	SAFETY	.doc	.wpd	.txt
Site Preparation and Clearing	SITEPREP	.doc	.wpd	.txt
Site Excavation	SITEEXC	.doc	.wpd	.txt
Site Development	SITEDEV	.doc	.wpd	.txt
Site Landscape	SITELAND	.doc	.wpd	.txt
Site Hardscape	SITEHARD	.doc	.wpd	.txt
Concrete Flat Work	CONCFLAT	.doc	.wpd	.txt
Concrete Foundations	CONCFOUN	.doc	.wpd	.txt
Masonry—Brick	BRICK	.doc	.wpd	.txt
Masonry—Concrete Masonry Units	CMU	.doc	.wpd	.txt
Metal—Railing	METRAIL	.doc	.wpd	.txt
Framing—Light-Gauge Metal	METALFRM	.doc	.wpd	.txt
Framing—Rough Carpentry	ROUGHFRM	.doc	.wpd	.txt
Carpentry—Siding	CARPSIDE	.doc	.wpd	.txt
Carpentry—Finish	CARPFIN	.doc	.wpd	.txt
Insulation—Batt/Blown	INSULBAT	.doc	.wpd	.txt
Roofing—Shingle	ROOFSHIN	.doc	.wpd	.txt

Sample Form	Filename	File Extension in		
		Word	WordPerfect	ASCII
Roofing—Built-Up	ROOFBLT	.doc	.wpd	.txt
Glazing	GLAZING	.doc	.wpd	.txt
Finishes—Drywall	DRYWFIN	.doc	.wpd	.txt
Finishes—Paint	PAINTFIN	.doc	.wpd	.txt
Finishes—Ceramic Tile	TILEFIN	.doc	.wpd	.txt
Finishes—Cementitious Stucco	STUCCFIN	.doc	.wpd	.txt
Finishes—Synthetic Stucco (EIFS)	EIFSFIN	.doc	.wpd	.txt
Finishes—Custom Millwork Wood Stairs	STAIRFIN	.doc	.wpd	.txt
Finishes—Cabinetry	CABFIN	.doc	.wpd	.txt
Finishes—Interior Cleanup	CLEANUP	.doc	.wpd	.txt
Flooring—Carpet, Vinyl, and Wood	FLOOR	.doc	.wpd	.txt
Appliances	APPLIANC	.doc	.wpd	.txt
Prefabricated Metal Fireplace	FIREMET	.doc	.wpd	.txt
Heating, Ventilating, and Air Conditioning	HVAC	.doc	.wpd	.txt
Fire Protection	FIREPROT	.doc	.wpd	.txt
Plumbing	PLUMBING	.doc	.wpd	.txt
Electrical	ELECTRIC	.doc	.wpd	.txt

Glossary

Arbitration—A process in which the parties to a dispute submit their cases to a neutral party for final and binding resolution.

Change order—A modification issued to a contract changing some aspect of the scope of work and usually the contract value.

Cost-plus contract—A contract that is awarded to reimburse the contractor for all costs incurred in executing the contract plus payment of some type of fee.

Contracting strategy—Identification of work packages to be contracted and selection of individual work packages or combinations of work packages for each trade contract.

Final lien release—A lien release provided by the trade contractor to the builder at the completion of all phases of the total scope of work. Builders should require that the final lien release be submitted with the final invoice for payment.

Indemnification—Shifting the financial loss from the party required to pay to the party that caused the loss. For example, builders generally write contracts requiring contractors to pay for any losses caused by them or their agents.

Liability insurance—Insurance that protects the insured against claims for injury or damage from third parties.

Lien—A legal claim for payment filed against privately owned real property for labor or material furnished for improvement of the property.

Lien release—A statement signed by the trade contractor that he or she will not file any liens against the project on account of labor, material, or equipment furnished pursuant to the contract.

Lump sum contract—A contract that provides a specific price for a specific scope of work.

Partial lien release—A lien release submitted by the trade contractor to the builder at the completion of each phase of the scope of work. Builders should require that a partial lien release be submitted with the invoice as each phase of work is completed.

Payment bond—A bond provided by the contractor guaranteeing that all employees, subcontractors, and suppliers will be paid.

Performance bond—A bond provided by the contractor guaranteeing that the contract scope of work will be completed as agreed in the contract.

Prequalification of contractors—Evaluating prospective contractors against a set of criteria to select those who will be invited to submit quotes for the work.

Request for information (RFI)—A written request from a trade contractor to a builder for clarification regarding the scope of work or the plans and specifications.

Request for quote—A document that asks contractors to submit a price for a detailed scope of work.

Retainage—A percentage of the contractor's payment that is retained by the builder until the entire project is completed.

Schedule of values—An allocation of the entire project cost to each of the various work packages required to complete the project.

Scope of work—A detailed description of the exact work to be performed by the contractor.

Shop drawings—Drawings prepared by the contractor to illustrate construction or installation details.

Subcontractor—A contractor who performs a portion of the scope of work assigned to another contractor.

Terms and conditions—Management requirements or procedures to be followed by the trade contractor when performing the contract scope of work.

Time and materials contract—A contract that reimburses the trade contractor for labor, material, and equipment costs and pays a fee for overhead and profit.

Trade contractor—Specialty contractors who perform trade work such as carpentry, plumbing, masonry, electrical, and mechanical tasks.

Unit price contract—A contract that contains an estimated quantity for each element of work and a unit price. The actual cost is determined once the work is completed and the total quantity of each work element is measured.

Warranty—A guarantee that all material furnished is new and is able to perform as specified and that all work is free from defects in material or workmanship. A trade contractor warranty should run concurrently with the builder's warranty starting at closing or the date the owner takes possession if owner built.

Workers' compensation insurance—Insurance designed to help employers meet their obligations under workers' compensation laws (e.g., if employees or trade contractors are injured on the job or suffer a job-related illness). Some employers are required by law to carry workers' compensation insurance.

Work breakdown—The process of subdividing the scope of work for the entire project into individual work packages.

Work package—A defined segment of the work required to complete a project. It may involve multiple project tasks or activities.

Related Publications Available from NAHB

All the publications listed below are available by calling the Home Builder Bookstore at (800) 223-2665, sending a request by Fax to (202) 822-0512, or sending an e-mail request to: *bookstore@nahb.com.* Home Builder Bookstore accepts Visa, MasterCard, and American Express. NAHB members receive a 20 percent discount on publications purchased through the Home Builder Bookstore. To receive the member discount, please enclose your member number with your order.

1. *Basic Construction Management: The Superintendent's Job,* 3d Edition, by Leon Rogers. Today's construction projects are more complex than ever. New construction managers can use this NAHB bestseller as a great training tool and more experienced superintendents can use the book to brush up on the latest techniques and technologies. From Home Builder Press. $27.50.

2. *Bar Chart Scheduling for Residential Construction,* by Thomas A. Love. Prepared specifically for small-volume builders, site superintendents, and residential remodelers, this book describes tried-and-true scheduling methods using simple bar charts. The diskette packaged with the book contains sample template files for bar charts created in Microsoft Excel and can be used with Excel or compatible spreadsheet programs. From Home Builder Press. $30.00.

3. *Contracts and Liability for Builders and Remodelers,* by David Jaffe. Now with diskette. One of the most requested books in the industry, the latest edition contains expanded material on increases in material costs, scheduling responsibilities, paying for inspections, mechanics' liens, owner occupancy before final inspection, updates on radon legislation and lead regulations, and more. From Home Builder Press. $35.00.

4. *Daily Field Guide: A Logbook for Home Builders,* by Tom Hrin. Designed by a working superintendent to keep multiple projects on schedule, this concise Field Guide also helps builders and superintendents record what happens on each jobsite during the course of construction. The logbook includes a calendar, conversions, and formulas cheat-sheet, phone chart for key contacts, projection chart, inspection chart, detailed project log for up to 15 projects, cleanup schedule, and weather log. From Home Builder Press. $37.50.

5. *Destination: Quality,* by Gilbert Veconi with Charles Layne. This chronicle of a small-volume home builder's quest for a higher level of customer service emphasizes a practical, manageable process for implementing a total quality management program, dispels some TQM myths, and includes an annotated resource list. From Home Builder Press. $28.00.

6. *How to Hire and Supervise Subcontractors,* by Bob Whitten. Learn the secrets to selecting, evaluating, and motivating subcontractors in this advice-packed book. Includes effective strategies to work with, rather than against, subcontractors; find subs who can do the job right; improve scheduling, maintain quality, control costs, and more. From Home Builder Press. $15.00.

7. *Model Safety & Health Program for the Building Industry,* 2d edition. With diskette. This model program provides builders a step-by-step process for setting up company safety administrative procedures, understanding requirements for governmental safety and health compliance, organizing training to meet company and field construction safety and health requirements, safety award and incentive programs, OSHA enforcement, and OSHA recordkeeping requirements. Model forms to assist in conducting a pre-job safety orientation, documenting safety violations and reprimands, investigating accidents, and conducting a construction safety audit are included. From NAHB's Labor, Safety and Health Services Department. $40.00.

8. *Production Checklist for Builders and Superintendents,* 2d edition, by John J. Haasl and Peter Kuchinsky. Now with diskette. Thousands of builders have used the Production Checklist as a construction management tool. This second edition covers safety, concrete production, rough trades, waterproofing, exterior stucco, finish trades, and final production, along with a typical construction schedule, sample pay schedules for subtrades, and a glossary of terms. All the checklists are included on a diskette that works with most word processing software. From Home Builder Press. $30.00.

9. *Quality Management: Best Practices for Home Builders,* by Edward Caldeira. This publication introduces the National Housing Quality Best Practices Series for home builders. With this first volume, the NAHB Research Center and Home Builder Press provide an executive summary of quality management, including an overview of four critical management areas: Company Vision, Strategic Management, Customer Focus, and Quality Management Systems. From Home Builder Press. $18.75.

10. *Residential Construction Performance Guidelines for Professional Bulders and Remodelers.* Bring customers' sometimes unrealistic expectations back down to earth with more than 240 guidelines in 12 major categories of new residential construction and remodeling. From NAHB Builder Business Services. $43.75.

11. *Scheduling Residential Construction for Builders and Remodelers,* by Thomas Love. This publication presents manual and computerized scheduling techniques that streamline the workload, reduce stress, and don't require 16-hour days. From Home Builder Press. $29.00.